THE HISTORY OF CHEMISTRY

The History of Chemistry

JOHN HUDSON
Anglia Polytechnic University, Cambridge

Chapman & Hall
New York

First published 1992 by
THE MACMILLAN PRESS LTD
Houndmills, Basingstoke, Hampshire RG21 2XS
and London
Companies and representatives
throughout the world

First published 1992 in North America by
Chapman & Hall, an imprint of
ROUTLEDGE, CHAPMAN & HALL, INC.
29 West 35th Street, New York, N.Y. 10001

Reprinted 1993

Library of Congress Cataloging-in-Publication Data
Hudson, John, 1943–
The history of chemistry/John Hudson.
p. cm.
Includes bibliographical references and index.
ISBN 0-412-03641-X. — ISBN 0-412-03651-7 (pbk.)
1. Chemistry—History. I. Title.
QD11.H84 1992   92–8311
540′.9—dc20   CIP

Printed in Hong Kong

Contents

To Judith, Simon, Mark and Benjamin

Preface

This book is written as a result of a personal conviction of the value of incorporating historical material into the teaching of chemistry, both at school and undergraduate level. Indeed, it is highly desirable that an undergraduate course in chemistry incorporates a separate module on the history of chemistry. This book is therefore aimed at teachers and students of chemistry, and it will also appeal to practising chemists.

While the last 25 years has seen the appearance of a large number of specialist scholarly publications on the history of chemistry, there has been little written in the way of an introductory overview of the subject. This book fills that gap. It incorporates some of the results of recent research, and the text is illustrated throughout. Clearly, a book of this length has to be highly selective in its coverage, but it describes the themes and personalities which in the author's opinion have been of greatest importance in the development of the subject.

The famous American historian of science, Henry Guerlac, wrote: 'It is the central business of the historian of science to reconstruct the story of the acquisition of this knowledge and the refinement of its method or methods, and—perhaps above all—to study science as a human activity and learn how it arose, how it developed and expanded, and how it has influenced or been influenced by man's material, intellectual, and even spiritual aspirations' (Guerlac, 1977). This book attempts to describe the development of chemistry in these terms.

The simple fact that science is a human activity is all too easily overlooked in the way science is taught to the student of today. This book not only presents an account of the development of chemistry as a scientific discipline, but gives brief biographical sketches of the chemists who have been most prominent in shaping the subject.

That we live in a world shaped to a considerable extent by the activities of the chemist is not in doubt. Some consideration is therefore given to the development of the chemical industry, the impact that manufactured chemicals have had on our world, and the organisation of chemistry within society.

When describing the history of any branch of science, there is a danger of concentrating on those ideas which seem to be forerunners of our modern views and giving less attention to those which, with the benefit of hindsight, seem to be in 'error'. It is important to realise that in any age scientists ask questions which seem important and relevant to them, and propose solutions that make sense at the time. We shall attempt to describe the activities of chemists and the theories they proposed throughout the evolution of the subject. Such a study is one of the best ways of learning about the process of science. By studying the history of chemistry, we are reminded that today's theories may be modified or discarded, just as so many ideas in the past have been changed or abandoned. Perhaps the ultimate reason for studying the history of chemistry is that we may thereby become better chemists.

Originally this book was to have been a collaborative venture between myself and Paul Farnham of the College of the Redwoods, in California. For various reasons, Paul was eventually unable to participate, but the project owes something to his enthusiasm in the early stages. I am indebted to Mr Alec Campbell of the University of Newcastle upon Tyne and Professor Mike Davies of Anglia Polytechnic University for reading the manuscript and making valuable suggestions, and also to Dr John Waterhouse of Anglia Polytechnic University for advice on aspects of the history of organic chemistry. I should also like to thank Professor Aaron J. Ihde for assistance with some of the illustrations.

No one writes a book in his spare time without inconveniencing his family, and my thanks are due to my wife and sons for their patience over the last three years.

Cambridge, 1992   J. H.

Reference

Guerlac, H., *Essays and Papers in the History of Modern Science*, p. 20 (Johns Hopkins University Press: Baltimore, 1977)

Acknowledgements

The author and publishers wish to thank the following for permission to reproduce illustrations:

The Ashmolean Museum, Oxford for 3.2. The Niels Bohr Archive for 11.12 (copyright). The British Library for 2.11. The Trustees of the British Museum for 1.1. The Cavendish Laboratory, University of Cambridge for 11.2, 11.8 and 11.11. Constable and Co. Ltd for 13.2 (from D. McKie, *Antoine Lavoisier*, 1952). Dover Publications Inc. for 2.3, 2.6 and 2.9 (from E. J. Holmyard, *Alchemy*; 1990); for 2.7 (from G. Agricola, *De Re Metallica*, translated by H. C. and L. H. Hoover, 1950). The Edgar Fahs Smith Collection, Special Collections Department, Van Pelt–Dietrich Library Center, University of Pennsylvania for 2.10, 6.1, 8.2, 8.3, 9.1, 9.6, 10.12 and 10.19. The Hulton–Deutsch Collection for 13.3. Prof. Aaron J. Ihde for 6.4 and 11.16. *Journal of Chemical Education* for 11.6 and 11.7 (from M. E. Weeks and H. M. Leicester, *Discovery of the Elements*, 1970); for 11.10 (from *J. Chem. Ed.*, 1980, **57**, 489). Lund University Library for 12.4 (from G. B. Kauffman *Inorganic Coordination Compounds*, Copyright 1981 Heyden, reprinted by permission of John Wiley and Sons, Ltd). Macmillan and Co. Ltd for 3.1 and 4.7 (from J. R. Partington, *A History of Chemistry* 1961–70); for 8.4 (from J. R. Partington, *A Short History of Chemistry*, 1948). The Mansell Collection for 1.7. The Metropolitan Museum of Art for 5.1 (Purchase, Mr and Mrs Charles Wrightman Gift, in honour of Everett Fahy, 1977. (1977.10), All Rights Reserved). Museum of the History of Science, Oxford for 11.9. The National Historical Museum, Frederiksborg, Denmark for 12.6. J. B. Pritchard, University Museum, University of Pennsylvania for 1.2. Ann Ronan Picture Library for 13.8. The Royal Institution for 7.2 and 7.5. The Royal Society for 4.4. The Royal Society of Chemistry for 10.10, 10.22 and 15.1. The Trustees of the Science Museum for 2.5, 5.3, 7.4, 11.3 and 14.5. Prof. Ferenc Szabadváry for 13.7, 14.1, 14.2, 14.3, 14.4 and 14.6 (from *History of Analytical Chemistry*, Pergamon, 1966). Ullstein Bilderdienst for 11.4.

In addition, they also acknowledge the following sources of further reproduced material:

1.5 P. Efstathiadis and Sons (from O. Taplin, *Greek Fire*, Jonathan Cape, 1989). 2.1 F. S. Taylor, *The Alchemists* (Heinemann, 1951). 2.4, 2.8, 3.3, 3.4, 4.3, 4.5 and 5.6 E. J. Holmyard, *Makers of Chemistry* (Oxford University Press, 1931). 4.1 *Mémoires de l'académie royale des sciences* (1718); 4.2 S. Hales, *Vegetable Staticks* (1727) (both from H. M. Leicester, *The Historical Background of Chemistry*, Dover, 1971). 4.6 F. W. Gibbs, *Joseph Priestley* (Nelson, 1965). 5.2 N. Powell, *Alchemy, The Ancient Science* (Aldus, 1976). 5.4 and 13.1 J. B. Dumas (ed.), *Oeuvres de Lavoisier* (1862); 5.5 A. L. Lavoisier, *Traité Élementaire de Chimie (1789); 7.1 Philosophical Transactions of the Royal Society* (1800); 9.2 *Annalen der Physic* (1860); 9.3 *Chemical News* (1866); 9.7 *Philosophical Magazine* (1882); 9.8 *Annalen der Chemie* (1873); 10.2 *Proceedings of the Royal Institution* (1865) (by courtesy of the Royal Institution); 10.3 F. A. Kekulé, *Lehrbuch der organischen Chemie*, Vol. II (1866) and *Annalen der Chemie* (1866); 10.6 *Annalen der Chemie* (1872); 10.14 Van't Hoff, *La Chimie dans l'Espace* (1875) (all from A. J. Ihde, *The Development of Modern Chemistry*, Harper and Row, 1964). 6.2 J. Dalton, *A New System of Chemical Philosophy* (1808); 6.3 *Philosophical Transactions of the Royal Society* (1813) (both from D. M. Knight (ed.), *Classical Scientific Papers, Chemistry*, First Series, Unwin Hyman of Harpercollins Publishers Limited, 1968). 8.1 J. S. Muspratt, *Chemistry, Theoretical, Practical, and Analytical* (1857–60) (from M. E. Weeks and H. M. Leicester, Discovery of the Elements, *Journal of Chemical Education*, 1970). 9.4 *Quarterly Journal of Science* (1864); 11.1 *Nature* (1879) (both from D. M. Knight (ed.), *Classical Scientific Papers, Chemistry*, Second Series, Unwin Hyman of Harpercollins Publishers Limited, 1970). 9.5 H. M. Leicester and H. S. Klickstein (eds), *Source Book in Chemistry* 1400–190, (McGraw–Hill, 1952). 9.9 W. Ramsay, *The Gases of the Atmosphere* (Macmillan, 1902). 11.5 and 11.23 The Bettman Archive Inc. 11.15 G. N. Lewis, *Valence and the Structure of Atoms and Molecules* (The Chemical Catalog Company, 1923. Reprinted by permission, The American Chemical Society). 11.18 I. Langmuir, *Journal of the American Chemical Society*, 1919, **41**, pp. 894 and 903. 12.3 National Academy of Sciences, *Biographical Memoirs*, 1902, **4**, 202; 12.8 Springer Verlag (both from G. B. Kauffman, *Inorganic Coordination Compounds*, Copyright 1981 Heyden, Reprinted by permission of John Wiley and Sons, Ltd); 8.7 Institut für Gerbereichemie Darmstadt; 15.2 The Society of Dyers and Colourists; 8.9, 10.13 and 13.6 (all from J. Kendall, *Great Discoveries by Young Chemists*, Nelson, 1953); 13.9 A. C. Barrington Brown (from H. F. Judson, *The Eighth Day of Creation*, Jonathan Cape, 1979).

Every effort has been made to obtain the necessary permissions with reference to copyright material. Should there be any omissions in this respect we apologise and shall be pleased to make the appropriate acknowledgements in any future edition.

1
Early Processes and Theories

The science of chemistry is concerned with the study of materials and the changes which they can be made to undergo. People started to transform natural materials into new ones more suitable for their purposes in Neolithic times. Although the science of chemistry as we understand it is of relatively recent origin, the early chemical technologists built up a fund of knowledge about chemical change. This was utilised by those who formulated the first theories about matter and the transformations it was observed to undergo.

THE NEOLITHIC REVOLUTION

A turning point in the history of humankind started to occur around 6000 BC when the fertile river valleys of the Nile, the Euphrates and the Indus were settled on a permanent basis. This happened as a consequence of global warming at the end of the last Ice Age, which resulted in the formation of large areas of desert. Peoples who had formerly been hunter–gatherers were now forced to live in the valleys where they began to tend crops and domesticate animals. This change of lifestyle, termed the *Neolithic revolution*, resulted in a large increase in population.

With this more settled pattern of existence, several crafts eventually emerged, which we now consider to be chemical. Perhaps of greatest significance was the smelting of metals, and we speak of the Stone Age being succeeded by the Bronze Age, which was in turn followed by the Iron Age. Other important crafts were making pottery and glass, tanning leather, the extraction of dyes and brewing. These new techniques, along with some others which were non–chemical in nature (such as spinning and weaving), contributed enormously to the advance of humankind. As the population expanded, cities were founded where the new crafts were practised and which acted as centres for trade. It was in these urban communities that the first scientific speculations occurred.

Fire was employed in many of the early chemical technologies. Fire had been tamed several hundred thousand years before the Neolithic revolution. It is conjectured that initially people merely tended fire captured from blazes started by

1

lightning strikes. This in itself was a huge achievement, for higher animals appear to have an instinctive fear of fire. Primitive people would have used their captured fire for warmth, for protection from wild animals and for cooking meat. The ability to initiate combustion presumably came later. The chemical reaction of combustion was destined to play a central role in the development of modern chemistry.

Our knowledge of the early chemical technologies is inevitably scanty. Some evidence is provided by the archaeologist. Sometimes an excavation may reveal the actual location at which pottery was made or a metal was smelted, but more often evidence of these processes is provided by the end–products that are unearthed. Chemical technologies had been practised for thousands of years before any written accounts were produced that have survived. The Roman writers Vitruvius (*fl.* first century BC) and Pliny the Elder (AD 23–79) describe many contemporary processes, but we believe that some of these had been carried out virtually unchanged for a very long time.

THE EXTRACTION OF METALS

The first metals to be used by man needed no extraction as they were found uncombined. Gold and occasionally silver and copper occur in this state. Iron is also found uncombined in some meteorites, and the iron in some early tools is known to be of meteoritic origin on account of the 8 per cent nickel which it contains. In 1894 when Robert E. Peary was exploring Greenland, an Eskimo showed him a huge meteorite that had acted as a source of iron for his tribe for the previous 100 years. Peary found that the meteorite still contained about 37 tons of iron.

The first metal to be obtained from an ore was copper, but we do not know when or where this chemical process was first performed. We can speculate that somehow a piece of the ore malachite (basic copper carbonate) was accidentally heated with charcoal, perhaps when a lump of malachite was used as one of the stones in a fire ring. By 4000 BC both copper and lead were in use in Egypt and Mesopotamia. By 3000 BC both regions were producing bronze, probably by smelting copper and tin ores together.

Many lead ores also contain silver, and between 3000 and 2500 BC the *cupellation* process was introduced to obtain pure silver from the lead–silver alloy which resulted from the smelting of such ores. The alloy was melted in a bone–ash crucible (the *cupel*) and the lead oxidised by a blast of air. The lead oxide was absorbed by the crucible, leaving a bead of silver. Cupellation was also used to refine gold. The impure gold was fused with lead in the cupel, and the impurities were removed with the lead oxide.

The extraction of iron became widespread much later, presumably because a much higher temperature is required to produce workable iron on account of its relatively high melting point. A furnace into which air is fed by bellows has to be used (Figure 1.1). It is not until 1000 BC that iron artifacts become common in the archaeological record.

Figure 1.1 Iron smelting depicted on a Greek vase

POTTERY AND GLASS

In palaeolithic times, meat was roasted over a fire using a spit, but once primitive peoples had adopted a settled mode of life and started to grow cereals and pulses, there was a need for cooking vessels. There must have been many occasions when someone noticed that the fire had hardened the patch of clay chosen for a hearth. The early potters would have learned to select suitable clays, to moisten them and to work them to a suitable consistency, to form them into vessels, and then to fire them. Early firings would have been performed on open fires, and the temperatures attained (450–750°C) would have done little more than dry out the clay. The invention of the kiln enabled higher firing temperatures (up to 1000°C) to be achieved. These cause chemical changes in the clay, making the pottery stronger and less porous. The earliest kilns date from around 3000 BC.

The discovery that a mixture of soda, sand and lime can be fused into a vitreous fluid, which on cooling results in a glass, is also very ancient. Glazed stone beads were produced in Egypt around 4000 BC. The addition of certain minerals to the fusion mixture produces coloured glasses, and these were used to glaze pottery and to make imitation gemstones. Glass was produced in considerable quantity in Egypt by 1350 BC. Some objects were cast in moulds, but glass vessels were made by dipping into molten glass a core of sand tied with a cloth and attached to a metal rod. The vessel was rolled into shape on a stone bench and the sand core was removed after cooling. The technique of glass blowing was invented about 50 BC and rapidly replaced the sand core technique.

PIGMENTS AND DYES

Paintings up to 30,000 years old have been found in caves in southern France. These were executed with natural mineral pigments; red with iron oxide, yellow with iron carbonate, and black with manganese dioxide (or sometimes soot). The Egyptians extended their range of pigments by using some substances prepared chemically, such as red lead oxide, Pb_3O_4, which they made by heating lead with basic lead carbonate.

The desire of people to colour their clothes resulted in the development of some chemical processes. It was known in ancient times that to render a dye fast the fabric must first be treated with a mordant to bind the dye to the cloth, and that the alums provided excellent mordants. Alums occur naturally quite widely, but they often contain iron salts, which would introduce an unwanted colour in the dyeing process. Although direct evidence is lacking, it would seem likely that even in ancient times alum was purified by a recrystallisation process.

The dyes used were of vegetable or animal origin. The madder plant provided a red extract and weld a yellow. Indigo was obtained either from the indigo plant or from woad. The plant was mashed in water and the mixture allowed to ferment, and after a time a precipitate of the blue indigotin dye was formed. This was collected and dried, and the dye was usually sold in this form. The indigotin was then reduced to a soluble colourless compound by an empirical chemical treatment (e.g. honey and lime were used in the Middle Ages), the fabric was then immersed in the solution and the material allowed to dry. The blue colour developed as a result of aerial oxidation during the drying process. This kind of procedure, in which an insoluble dye is formed in the cloth after it has been soaked in a soluble precursor, is today known as *vat dyeing*.

The most prized dye of antiquity was Tyrian purple (or royal purple). We now know that this dye is dibromoindigotin. The ancients obtained an almost colourless precursor, a few drops at a time, from the glands of a shellfish that they harvested from the Mediterranean (Figure 1.2). The purple developed when the treated cloth was dried in the sun. Other species of shellfish yielded a violet colour, which was due to a mixture of indigotin and dibromoindigotin. It is clear that this kind of dyeing was a very considerable industry, for huge deposits of shells have been found, all broken at the appropriate place to obtain the extract from the gland. With the process being carried out on such a scale, there was a need to reduce the coloured compounds that had been formed prior to the dyeing operation. Pliny describes how the extract was boiled in a vessel of *plumbum*. This is presumably tin (*plumbum album*), as indigotin and dibromoindigotin are reduced to a colourless soluble compound by tin and alkali.

Figure 1.2 A thirteenth century BC spouted vat from Sarepta, Lebanon. Traces of dibromoindigotin were found inside. The vessel could have been used to separate the solution of the dye precursor from the crushed molluscs

Other colouring materials used since the earliest times include a series of scarlet compounds extracted from insects living on certain species of oak. This source was used in Europe until the sixteenth century, when it was superseded by the cochineal insect introduced by the Spaniards from the New World.

EARLY SPECULATIONS ON THE NATURE OF MATTER AND ITS TRANSFORMATIONS

To the early scientific philosophers the production of a metal such as copper from malachite in a charcoal fire must have seemed almost magical, and the extraction of metals may have been performed in temple precincts where priests could supervise the skilled artisans performing their tasks. The priests may well have assisted in the process, as they thought, by ensuring that the astrological signs were favourable and that the gods were propitiated. In any event, the priests were the first group to have an intimate knowledge of the chemical technologies of the day and also the leisure to speculate on the nature of the changes that occurred.

The earliest theory to explain the bewildering variety of substances in the world was that everything was created from a first principle. The Babylonians took this to be water. One Babylonian legend tells how in the beginning there was only sea, and that dry land had been created from sea by the god Marduk. The idea of a first principle from which all other materials are derived was used in many cultures as a basis of the explanation of chemical change.

The Babylonians and Egyptians made careful observations of the night sky. Because there seemed to be a correspondence between the positions of the heavenly bodies and important events on earth, such as the annual inundation of the Nile, it seemed that the stars and planets exercised a power that controlled events on earth. Studying the heavens was therefore both a religious duty and a matter of practical utility. The idea also arose that the sun, moon and five planets were associated with the metals.

Babylonian and Assyrian thought tended to be more astrological than that of the Egyptians, but ideas were exchanged between these early civilisations for over 2000 years and furnished concepts that the Greeks were able to develop and extend.

IONIAN SCIENCE

The thinkers of classical Greece sought knowledge for its own sake, and attempted to formulate a comprehensive philosophy encompassing all aspects of the material world. The early period of Greek thought, between 600 and 500 BC, is sometimes called the *Ionian*. Ionia was the name the Greeks gave to the Mediterranean coast of Asia Minor, where they established several colonies. It was in the Ionian colony of Miletus that philosophical speculation started, and it continued in centres such as Ephesus and on the island of Kos.

The principal philosophers of this early period were Thales (*fl.* 585 BC), Anaximander (*fl.* 555 BC), Anaximenes (*fl.* 535 BC), and Heraclitus (*fl.* 500 BC). They sought a materialistic explanation for the universe, and made no reference to any supernatural agency. Their theories were founded on observations both of nature and of technical processes, but they performed no systematic experimental work. In their attempts to achieve a comprehensive view of the world, they produced bold and sweeping generalisations.

Thales upheld that everything in the material world was a single reality appearing in different forms, and that this single reality was water. He thought water was converted into other substances by natural processes, like the silting up of the delta of the Nile in which water was apparently converted into mud.

Anaximander attempted a more comprehensive description of the universe, but took as his first principle a mysterious primitive substance which he named the *Boundless* and from which all other materials, including water, were generated. Anaximenes rejected Anaximander's mysterious primitive substance and held that the first principle was mist. He maintained that mist could be rarefied to yield fire, and could also be condensed to produce water and earth. Anaximenes imagined that these processes were occurring all the time, and that everything in the world was in a constant state of change.

Heraclitus expanded on this concept of continual change. His primary material was fire, which could condense to water and then to earth. He associated opposite qualities with materials. Thus he associated the qualities of hot and cold with fire and water, and the moist and dry qualities with mist and earth.

THE WESTERN GREEK PHILOSOPHERS

Towards the end of the sixth century BC another tradition in Greek philosophy was founded by Pythagoras on the western edge of the Greek world. Pythagoras was born about 560 BC on the Ionian island of Samos and when he was about 30 years old he migrated to Croton in southern Italy, where he founded a religious cult based on mathematics. The Pythagoreans saw in mathematics a key to understanding the universe and also a means of purifying the soul. Instead of the materialistic philosophy of the Ionian school, the Pythagoreans attempted to account for phenomena in terms of mathematical relations. An area in which this approach was successful was music, where the Pythagoreans discovered that two vibrating strings whose lengths were in simple numerical ratio produced harmonious sounds. The Pythagoreans were the first to attempt to describe phenomena in quantitative mathematical terms.

Even further from the materialistic philosophy of the Ionian school was Parmenides, who came from Elea in southern Italy and lived around 500 BC. He believed that truth should be sought by reason alone and that the evidence of the senses should not be trusted. He therefore rejected observation and maintained that all change is illusory. He thus adopted a position diametrically opposed to his Ionian contemporary Heraclitus. In no sense can Parmenides be called a scientist, but he is important because later philosophers had to consider his arguments.

The next major philosopher among the western Greeks was Empedocles of Agrigentum in Sicily, who was active around 450 BC. He rejected Parmenides's philosophy, and like the Ionians based his views on observation. He is famous for his experiment with the *clepsydra*, or water clock. This was a conical vessel with a small hole at the point of the cone. The cone was filled with water, and time was measured by the emptying of the vessel. Empedocles inverted the clepsydra, lowered it into a bowl of water and noticed that if he placed his finger over the small hole then water did not enter the larger opening. He thus provided evidence that, although air could not be seen, it was a material substance. This was a conclusion of tremendous importance, and helped the Greek atomists to explain the properties of all materials in terms of particles which were too small to be seen.

Empedocles proposed that there were four unchanging substances: earth, air, fire and water, which were the *roots* of all things. He also maintained that there were two basic forces, namely love and hate, or attraction and repulsion, which were responsible for combining and separating the roots. He proposed that different substances are formed by the roots combining in different proportions, and he thus demonstrated that a vast number of different substances could exist. The roots were not to be likened to the familiar substances that go by the names earth, air, fire and water, but were to be identified with their essential characteristics.

SCIENCE ON THE GREEK MAINLAND

After 500 BC important theories were advanced by philosophers who lived in mainland Greece. Anaxagoras (500–428 BC) came from Ionia but settled in Athens. He proposed a completely different explanation of the composition of materials and of chemical change, maintaining that in every material there is a portion of every other material. Thus wheat contains not only wheat but also hair, skin, bone, gold, etc. When wheat is digested, some of the skin separates and joins to the skin already present in the animal. However, complete separation is never achieved, and every material always retains a portion of every other material.

Leucippos also came from Ionia, and settled at Abdera on the Greek mainland around 478 BC. He was the first person to propose an atomic theory of matter, but no portion of any of his work survives. Our knowledge of the Greek atomic theory comes from his pupil Democritos, who also came from Abdera and lived around 420 BC. Democritos maintained that all the materials in the world were composed of indivisible *atoms* (from the Greek *atomon* meaning indivisible). Atoms of different shape, arranged and positioned differently relative to each other, accounted for the different materials of the world. Atoms were in random perpetual movement in a *void*. Democritos also explained the feel and taste of substances in a new way. These were not inherent qualities of the material but a result of the effect the atoms of the material had on the atoms of our sense organs.

The atomic theory provided an immediate explanation for changes which we now regard as chemical; the new material was formed as a result of a rearrangement of the atoms. The theory was to be rejected by Aristotle, and, because for almost 2000 years after his death he was regarded as the supreme authority on scientific matters, atomism made little impact on science until relatively recent times.

To some extent the theories of Empedocles, Anaxagoras and the atomists were all attempts to accommodate Parmenides's assertion that change is impossible. In all these theories the fundamental constituents of matter remain unchanged. Chemical change is a manifestation of a change in the proportions of the roots (Empedocles), or a result of the partial separation of materials (Anaxagoras), or a consequence of the rearrangement of the atoms (Leucippos and Democritos).

One of the most famous of the Greek philosophers was Socrates (470–399 BC). The *Socratic method* is the name given to the process by which an argument starts from an incontrovertible statement based on common experience and is developed by means of clearly defined logical rules. Socrates was not concerned with the nature of matter or the structure of the universe; his philosophy was concerned with moral and ethical issues. His aim was to deduce rules that would enable people to be good citizens and live together in peace.

The tradition founded by Socrates was of considerable importance in the subsequent development of mathematics, but science as we know it requires a different approach. It was not until the sixteenth century that the modern scientific method was developed. The science of the intervening period was influenced by Socrates's pupil Plato (c. 427–347 BC), and above all by Plato's pupil Aristotle (384–322 BC).

Plato was influenced both by Socrates and by the Pythagoreans. The Pythagoreans had discovered that there could only be five regular polyhedra with all their sides made up of identical regular polygons (Figure 1.3). Plato identified the four roots of Empedocles with four of the five regular polyhedra. Thus he identified fire with the tetrahedron (sharp and spiky), air with the octahedron, water with the icosahedron (almost spherical and slippery) and earth with the cube. The twelve–sided dodecahedron was identified with the twelve signs of the zodiac.

Plato was now able to propose, in contrast to Empedocles, that some of the primary roots are interconvertible. The faces of the tetrahedron, octahedron and icosahedron are all equilateral triangles, and so Plato proposed that an icosahedron of water may be converted to two octahedra of air and one tetrahedron of fire. Curiously Plato imagined that this occurred through the rearrangement of the six small half–equilateral triangles into which a large equilateral triangle may be divided (Figure 1.4). However, the cubes of earth, having square faces constructed from right–angled isosceles triangles, were unable to be formed from, or converted into, any of the other polyhedra.

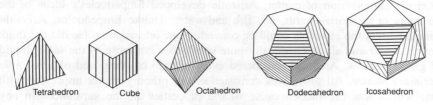

Tetrahedron Cube Octahedron Dodecahedron Icosahedron

Figure 1.3 The five regular polyhedra discovered by the Pythagoreans

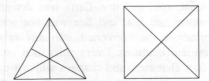

Figure 1.4 Plato's demonstration that an equilateral triangle can be divided into six small half–equilateral triangles, and that a square can be divided into four right–angled isosceles triangles

It is unclear to what extent Plato imagined the polyhedra comprising earth, air, fire and water to be solid bodies. In general Plato did not seek mechanical causes for natural phenomena, but explained events in terms of their eventual outcomes. He believed everything was ordered by a Divine Intelligence to produce the best of all possible worlds.

ARISTOTLE'S SCIENCE

The most famous of the Greek scientific philosophers was Aristotle. His interests were very wide, and his influence on subsequent thought was enormous. When

considering a subject Aristotle undertook a critical review of the theories proposed
by his predecessors before developing his own views.

> Aristotle (Figure 1.5) was born in Stagira in Macedon, where his father was
> physician to the Macedonian king. As a young man Aristotle went to Athens
> to study with Plato at the *Academy*, which was a public gymnasium outside
> Athens where young men argued and debated as they rested after exercising.
> After twenty years in Athens Aristotle spent about five years studying natural
> history, first at Assos on the Ionian coast and then on the island of Lesbos.
> Around 342 BC he returned to Macedonia to be tutor to Alexander, who was
> heir to the throne of Macedon and was destined to conquer most of the
> known world. After a period in his home town of Stagira, he returned to
> Athens in 335 BC and founded his own school at another gymnasium called
> the *Lyceum*. Alexander died in 323 BC, and in the face of growing anti-
> Macedonian feeling Aristotle went to the island of Euboea, where he died the
> following year.

On the composition of matter, Aristotle developed Empedocles's ideas of the
four roots or *elements*—earth, air, fire and water. Unlike Empedocles, Aristotle
maintained that the elements could be converted one into another. He did not think
of earth, water and air as chemically pure substances but more as the solid, liquid
and gaseous states. Aristotle considered everything to be composed of *proto hyle*
or *primary matter*. All substances consisted of this primary matter impressed with
form, which was the hidden cause of the properties of the substance. In any
transformation, the primary matter persisted unchanged but the form was altered.

Neither matter nor form could exist in isolation, and the simplest combinations
of matter and form were the elements. The elements could be analysed in terms of
four *qualities*: hot, moist, cold and dry. Earth was dry and cold, water was cold
and moist, air was moist and hot, and fire was hot and dry (Figure 1.6). One
element could, in principle, be converted into any other by the addition and
removal of the appropriate qualities. Every substance on earth was composed of
combinations of the four elements, and changes which we now call chemical were
explained by an alteration in the proportions of the four elements.

Aristotle made little attempt to produce evidence to support his four–element
theory. His later supporters pointed to the fact that, when a piece of wood is
burned, *fire* issues from it, *water* oozes out of it, *air* is produced (in the form of
smoke) and *earth* (ashes) remain behind. It had to be admitted that many other
substances could not be analysed in this way, but in the absence of a better theory
Aristotle's explanation of the material world remained in vogue until at least the
sixteenth century.

Aristotle's four–element theory was to exert a considerable influence on early
laboratory chemistry, or *alchemy* (Chapter 2). The alchemists were also influenced
by Aristotle's ideas on the formation of metals. He believed that metals and
minerals were formed from *exhalations*. The two exhalations were a moist
vaporous one, which was formed when the sun's rays fell on water, and a dry
smoky one, which arose from the land. When these exhalations became imprisoned
in the earth, they formed minerals or metals depending on which exhalation was

present in greater quantity. When the predominant exhalation was the dry smoky one, minerals were formed, and metals resulted when the moist vaporous exhalation was present in excess.

Figure 1.5 The modern statue of Aristotle (384–322 BC) at Stagira

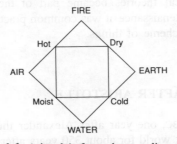

Figure 1.6 Aristotle's four–element diagram

Aristotle explained the motion of objects by saying that each element goes naturally to its proper place. Earth falls downwards as far as possible, water comes to rest above earth, air rises above water, and fire rises above air. The four elements were confined to the region below the moon. The fifth element was the material of the heavens, which had a circular motion.

Aristotle rejected the atomic theory of matter. Democritos had maintained that atoms were in perpetual motion in a void, and Aristotle criticised both the concepts of atomic motion and of the void. Aristotle accounted for motion in terms of the natural tendencies of bodies, and the perpetual random motion of atoms did not concur with his ideas. Aristotle rejected the idea of the void partly as a result of erroneous beliefs concerning falling bodies. He believed that the speed with which a body falls to earth is proportional to its weight; he had no understanding of the acceleration of falling bodies. From the observation that bodies fall more slowly in water than in air, Aristotle also concluded that the speed of fall was inversely proportional to the thickness or resistance of the medium. Since a vacuum would have a resistance of zero, the speed of fall of a body in it would be infinite. This was impossible, so Aristotle concluded that a vacuum could not exist.

While Aristotle's dynamics contained incorrect conclusions based on superficial observations, much of his zoology was sound. Some of his observations on marine animals were not confirmed until relatively recent times. An example is his report that after laying her eggs the female catfish swims away and leaves them to be guarded by the male. For many years this was thought to be incorrect, but in 1856 it was discovered that this is an accurate account of the behaviour of the particular species Aristotle was observing. Aristotle realised how well adapted were the animals he was studying to the environment in which they lived. He classified living things on the basis of increasing complexity, commencing with plants, sponges and jellyfish and ending with mammals and man. Aristotle had no concept of evolution; he imagined all species to be unchanging. Living things were distinguished from inanimate matter by the possession of a *soul*, and man had the most highly developed soul of all living things. Aristotle, like most Greek philosophers, speculated on the heavens and the earth's place in them. Aristotle believed that the universe was spherical with the earth fixed at the centre. The sun, moon, planets, and stars were held on clear crystal spheres surrounding the earth. As well as being extremely wide ranging, Aristotle's natural philosophy acquired an immense authority which remained almost unchallenged for nearly two thousand years. His cosmological theories became part of the doctrine of the Christian church, and until the Renaissance it was common practice to attempt to fit all new facts into Aristotle's scheme of things.

GREEK SCIENCE AFTER ARISTOTLE

Aristotle died in 322 BC, one year after Alexander the Great. The culture which flourished in the Greek world for about 350 years after Alexander's death is given the name *Hellenistic*. During the early part of this period Athens continued to be an important centre for scientific speculation, but it was not to be long before science started to flourish in the new city of Alexandria.

When Aristotle left Athens in 323 BC, Theophrastus (371–286 BC) took over as head of the Lyceum. Theophrastus attempted to continue Aristotle's study of the natural world by writing about topics which Aristotle had not had time to consider.

The most important areas of Theophrastus' work were botany and mineralogy. His mineralogical treatise (entitled *On Stones*) was to remain the leading work on the subject for nearly two thousand years. Like Aristotle, Theophrastus classified the materials found underground as either metals (in which water predominated), or stones (in which earth predominated). However Theophrastus went on to classify the stones according to properties such as colour, hardness, and the changes which occurred when they were heated strongly. He summed up the importance of the art of inducing chemical changes thus: 'Art imitates nature and also creates its own substances, some for use, some merely for their appearance ... and some for both purposes like quicksilver'.

When Theophrastus died in 286 BC Strato took over at the Lyceum, and he remained in charge until 268 BC. No writings of Strato have survived, so his ideas have to be inferred from the accounts of later writers. He departed from Aristotle's views in maintaining that falling bodies accelerate and that a void is possible. However the really important aspect of Strato's work is that it involved planned experiments, not mere observation. He described the construction of a hollow metal sphere connected to a metal pipe. The air inside the sphere could be compressed or rarefied by blowing or sucking, and Strato concluded that air was composed of particles that could either be pushed closer together or pulled further apart in the surrounding void. This is one of the first examples of a scientific experiment conducted with a purpose–built piece of apparatus.

Neither Theophrastus nor Strato put forward a wide–ranging theory of matter to rival Aristotle's. However at about the same time two schools of philosophy, the Epicureans and the Stoics, were founded which did produce comprehensive physical theories. For both the Epicureans and the Stoics the aim of philosophy was to secure human happiness. To this end the purpose of scientific enquiry was to explain the causes of natural phenomena such as lightning and earthquakes without reference to any kind of supernatural agency, and in that way people would be free from superstition and fear. However, although the two schools of philosophy attempted to understand natural phenomena for the same fundamental reason, the physical theories which they proposed were very different.

The Epicureans were named after Epicurus (341–270 BC) who set up his school in Athens around 300 BC. Epicurus claimed that his views owed nothing to earlier authorities, but he adopted an atomic hypothesis that was clearly based on that proposed by Leucippos and Democritos. Epicurus believed that the properties of materials were determined not only by the shape, arrangement and position of their atoms but also by their weight.

Once the Epicureans had suggested a reasonable explanation for a natural phenomenon, they took their enquiry no further. They were content to show that there was no need to involve any Deities (who might have to be propitiated). They were not troubled if two or more explanations could be advanced for a particular phenomenon, and they made no attempt to decide between such alternative theories. An account of Epicurean philosophy was given by the Roman Lucretius (c. 100–55 BC) in his poem *De Rerum Natura*. It was through Lucretius's work that Pierre Gassendi (1592–1655, Chapter 3) learnt of the Epicurean philosophy and became an early European advocate of the atomic theory of matter.

Whereas the Epicureans embraced the atomic theory of matter, the Stoics believed that matter was a continuum. The most important figure among the early Stoics was Zeno of Citium, who was a slightly younger contemporary of Epicurus. The Stoics held that matter is composed of two principles, the passive and the active. The active principle was called *pneuma* (breath or vital spirit), and it was the combination of pneuma with passive matter that resulted in a substance having particular properties. The Stoics also assigned to pneuma the property of holding together a complex structure such as a plant or an animal, and indeed the universe itself was supposed to be pervaded by a kind of pneuma. The Stoic idea of pneuma was to be taken up by some of the alchemists.

The Greeks performed little experimental work. Their scientific theories were vast edifices of speculation, which were sometimes based on observations of natural processes and existing technologies, and sometimes not. It has been suggested that the attitude of the Greek philosophers towards experimentation arose because all manual work was done by slaves, and that such activities were therefore demeaning to the elite to which the philosophers belonged. Certainly there is an element of this attitude in the writings of Plato, but a more likely explanation is that the idea of designing and performing experiments rarely occurred to philosophers, to whom abstract intellectual activity was the only worthwhile pastime. To us this seems a serious shortcoming, but we must remember that we are products of a modern scientific culture in which experimentation is an integral part of the scientific method.

THE RISE OF ALEXANDRIA

The Greek world had been dramatically changed by Aristotle's pupil, Alexander the Great (356–323 BC). Prior to Alexander, Greece had been composed of a number of small city states. Alexander not only united these city states but also conquered much of the known world. His empire included Egypt and Mesopotamia, and stretched eastwards towards India. He founded cities which acted as centres of Greek culture in the lands he had conquered. The most important of these cities in the history of chemistry was Alexandria, which was founded at the mouth of the Nile in Egypt in 332 BC.

The Hellenistic culture that flourished in Alexander's empire was distinct from that of classical Greece in that it was to some extent a fusion of the culture of Greece with that of the conquered lands. When Alexander died, his empire was split among his generals, and Ptolemy Soter (c. 266–283 BC) became ruler of Egypt. He founded a dynasty, which made Alexandria the most important centre of learning in the entire world. He built a vast library (Figure 1.7) together with the museum (or university).

Not only did science flourish in Alexandria, but it was also of a more applied and practical nature than its Greek precursor. The most famous Alexandrian scientist was Hero (AD 62–150), who invented a number of mechanical devices that utilised the pressure of air or steam. Among these were a simple steam turbine and a device by which the doors of a shrine were opened automatically when a fire was lit on the altar of the temple. Hero accepted the views of Strato and believed that gases consisted of particles in a void, and when a gas was compressed the particles were forced closer together. Other famous scientists and mathematicians of the Hellenistic period who studied or worked at Alexandria include Euclid, Ptolemy the astronomer, Eratosthenes and Archimedes.

Figure 1.7 A reconstruction of the library at Alexandria

It was in Alexandria that some of the earliest attempts at chemical experimenta-tion took place. These investigations were not conducted with the intention of learning more about the properties of materials or of making new substances. The principal object of early laboratory chemistry, or *alchemy*, was the artificial production of gold.

2

Alchemy

The first laboratory workers were alchemists, and their principal goal was the conversion of base metals into gold. Alchemy was first practised in Alexandria, and the Hellenistic phase of alchemy lasted until the seventh century AD, when the dramatic expansion of Islam occurred. For the next 500 years the chief practitioners of alchemy came from the world of Islam. Then, in the twelfth century, when Islamic power was in decline, Arabic alchemical manuscripts began to be translated into Latin, and the European phase of alchemy began. The art was practised in Europe for the next 500 years, before entering a slow decline.

Western alchemy was the prelude to modern chemistry, but alchemy was also practised in China from before 175 BC to around AD 1000. It is almost certain that Islamic alchemy incorporated some ideas that originated in China.

Chemistry is indebted to alchemy for the development of many laboratory techniques. With the aid of these techniques, some important substances were isolated, for example ethanol and the mineral acids. Although these discoveries were made at a time when alchemy was still practised, there is no evidence that they occurred in the course of a quest for gold. The only well–attested case of an important discovery being made during alchemical research was the isolation of phosphorus by Hennig Brand in 1669.

ALEXANDRIAN ALCHEMY

For many centuries before Alexandria was founded, Egypt had produced craftsmen whose skill in making objects of gold, silver and precious stones was unrivalled. The fabulous treasure found in the tomb of Tutankhamun, a relatively insignificant Pharaoh who died in 1352 BC at the age of eighteen, is evidence of this fact.

Egyptian craftsmen were also proficient at producing imitation gold and in giving articles a golden surface. Such techniques are somewhat similar in their effects to our modern practice of electroplating. In neither case is there necessarily any attempt to deceive.

Our knowledge of these technical processes of the Egyptians derives chiefly from two papyri found when a tomb was opened about 150 years ago. They are known as the Stockholm and Leyden papyri, after the cities where they are now kept. They date from the third century AD, but almost certainly describe processes that had been known for centuries. They provide recipes for the production of debased gold, for example by alloying it with copper, and also give methods of imitating various precious stones by preparing types of coloured glass. One interesting process is a method of giving an article a surface of pure gold. The article was made of adulterated gold and then heated with iron(II) sulphate, alum and salt. The acids released would have dissolved the base metals on the exterior of the object, thus leaving a surface of pure gold.

While there is no suggestion that the Egyptian metallurgists claimed that they were transmuting other metals into gold, it is likely that their processes made a deep impression on the scientific philosophers of Alexandria. The contemporary view was that there was a range of 'golds' of varying quality, for some native gold contains copper. In the light of Aristotle's four–element theory, which predicted that any material could be changed into any other, it was reasonable to attempt to improve the 'goldness' of a base metal. Furthermore, Aristotle had said that metals were formed when exhalations became imprisoned in the dry earth. There was a belief that metals slowly matured in the earth, and that the ultimate state of perfection was reached when the metal became gold. It might therefore be possible to accelerate this process artificially. Another important idea that the early alchemists borrowed from Greek philosophy was the Stoic concept of pneuma (Chapter 1).

It therefore seems likely that alchemy was born in Alexandria as a result of contact between philosophers in the Greek tradition and skilled Egyptian craftsmen. From an early date, alchemy was also subject to a number of mystical and religious influences. Two of these were Gnosticism and Neo–Platonism, which were in vogue during the first three centuries of the Christian era. Gnosticism contained elements of earlier Greek philosophy, and also borrowed some ideas from Christianity and other religions. Gnostics claimed that they possessed a secret knowledge or *gnosis*, which had been transmitted to them by occult means.

Neo–Platonism was founded by Plotinus (AD 205–270) and was an important system of philosophy and mysticism. The neo–Platonists were responsible for a collection of texts based on earlier Egyptian writings ascribed to the Egyptian god Thoth, whom the Greeks identified with their god Hermes. When these Hermetic writings of the neo–Platonists were discovered by the Europeans, they were initially thought to date from Egypt at the time of Moses, and their supposed author was referred to as *Hermes Trismegistos* (Chapter 3).

The Hermetic writings had an impact on Alexandrian alchemy and on later European alchemy and science. Among the ideas they contained were that certain combinations of numbers possessed mystical properties, and that the stars could exert an influence on human endeavours. The secrets of the magical world portrayed in the Hermetic writings were only open to the chosen few. It is because of the influence of the Hermetic writings that alchemy was often referred to as the *hermetic art*, and this is also the origin of the term *hermetically sealed*.

It is impossible to give a precise date to the dawn of alchemy. One of the earliest writers whom we can definitely identify is Zosimos, who worked in Alexandria around AD 300. He was preceded by about 100 years by the more

shadowy figure of another author writing under the name of Democritos. This was
not, of course, the Democritos of Abdera, who had described the atomic theory
some 600 years earlier, but it was common among alchemical writers to ascribe
their texts to famous predecessors in an attempt to give them an air of authority.
The probable author is Bolos, who came from Mendes near Alexandria. Both Bolos
and Zosimos undoubtedly drew on the earlier work of others, and the best estimate
we can give of the date when alchemy started is some time in the first century AD.

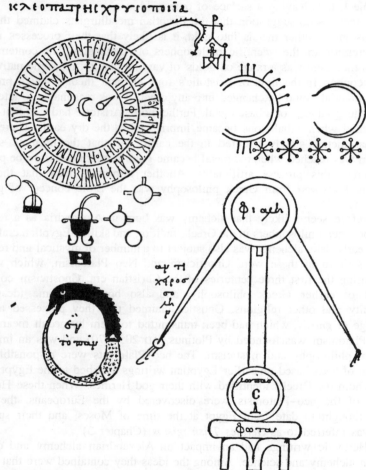

Figure 2.1 Alchemical symbolism from a manuscript from St Mark's in Venice. This
manuscript was transcribed in the tenth century from an earlier source, and is entitled 'The
goldmaking of Cleopatra'. The original author is believed to have written soon after the
beginning of the Christian era, and is not to be confused with the famous Queen Cleopatra.
The symbol at the bottom left is an Ouroboros, and at the bottom right is a distillation
apparatus. Immediately above the Ouroboros are other representations of apparatus, and the
circular design at the top left represents the unity of matter, and encloses the old symbols for
mercury, silver and gold. The symbols at the top right may represent the transmutation of a
base metal into a noble one

Zosimos is reputed to have composed a monumental work on alchemy, which ran to 28 books, and fragments attributed to him survive. Much of the writing is highly symbolic and obscure, and it would appear that Zosimos intended his work to be incomprehensible to the uninitiated, but to convey some meaning to the instructed. Not surprisingly, the meaning of much of Zosimos's writing eludes us today, but it is clear from some passages that he had a wide knowledge of practical chemistry.

Many alchemical works, even from the earliest period, contain illustrations intended to symbolise some aspect of the alchemical process (Figure 2.1). A recurring image is that of the *Ouroboros*, a serpent eating its own tail, and symbolising the essential unity and interconvertibility of matter. Also, from the earliest stage of alchemy derives the practice of symbolising metals by the signs of the planets to which it was imagined they were connected (Figure 2.2). Metals were indicated by such symbols until the early nineteenth century, when Dalton proposed his own symbols for elements in conjunction with his atomic theory.

Sun	☉	Gold
Moon	☾	Silver
Mercury	☿	Mercury
Venus	♀	Copper
Mars	♂	Iron
Jupiter	♃	Tin
Saturn	♄	Lead

Figure 2.2 The planetary symbols for the metals. These symbols were used from about AD 500. In earlier texts, mercury was given the symbol of the waning moon, and tin was given the symbol later ascribed to mercury

Reconstructing the procedures of the early alchemists is no easy task. In general, if an alchemist was to attempt to transmute a metal such as copper into gold, his first task, using the Aristotelian idea that substances were composed of matter impressed with form, was to remove the form of copper from his starting material. This was often portrayed as the death of the copper. The desired form of gold then had to be introduced, often in the form of a seed, and nurtured and allowed to grow. Gentle heat was often used at this stage in an attempt to introduce *pneuma*.

To the Alexandrians the colour of a metal was a fundamental property, and a material being subjected to an alchemical transmutation was expected to exhibit a certain sequence of colours. The first stage was blackening or *melanosis*, and represented the loss of colour when the starting material died. Then followed *leucosis* (whitening) and *xanthosis* (yellowing), which represented the production of gold. Sometimes a final stage is mentioned in the alchemical texts. This was *iosis*, and was the production of a purple iridescence on the surface of the gold.

In the first stage of the process, copper or lead could be blackened by conversion to the sulphide by direct reaction with sulphur. Orpiment (arsenic sulphide) was used to whiten copper, and calcium polysulphide solution (prepared by heating lime, sulphur and vinegar) was used to tinge the surface of a metal with a yellow colour. Presumably between the blackening and the whitening stage the black sulphide was smelted back to the metal. Clearly a great deal of speculation is involved in any modern attempt to reconstruct ancient alchemical processes, and it is important to remember that many reagents used by these early workers would have been very impure by modern standards.

It is tempting to suggest that on occasions an alchemist must have concluded that he had been partially successful. If the starting material was copper, and a zinc–bearing mineral was introduced at some stage in the process, the product could have been a type of brass. The yellow product must have seemed much nearer to gold than the starting material. Brass was known in the ancient world, although pure zinc was not isolated until the sixteenth century. Brass was made by roasting together calamine (zinc carbonate), charcoal and copper. To the ancients, brass was about six times more valuable than copper.

One interesting piece of apparatus invented by the Alexandrian alchemists was the *kerotakis* (Figure 2.3). One or more base metals were placed on the support, and sulphur was placed underneath. When the fire was started, the sulphur vapour attacked the metals and the sulphides would be carried downwards by the descending condensed sulphur. The sieve would catch pieces of unreacted metal. The resulting black mixture in the base was then heated in an open vessel to remove the excess sulphur and then subjected to the next stage of the process. The kerotakis, when used in this way, would clearly achieve the first stage (melanosis) in the desired sequence of colour changes. The lower part of the apparatus, in which the 'dead' product accumulated, was sometimes referred to as Hades. The kerotakis was also used to treat metals with the vapours of mercury and probably arsenic, although it is not certain that arsenic was isolated by the ancients.

Other laboratory operations performed by the Alexandrian alchemists included solution, filtration, crystallisation, sublimation, and distillation. The art of distillation shows a steady improvement over the years. In the earliest form of distillation apparatus, the vapours ascending from the flask or *bikos* were condensed in the hood or *ambix*, and then ran down the spout into the receiver (Figure 2.4). The Arabs later used the name *alembic* for the condensing hood. This method of condensation was inefficient, and it was not until much later that water-cooled condensers were introduced. The alchemists employed several different kinds of furnace, and also introduced the sand bath and water bath for heating purposes. The water bath is reputed to have been invented by one of Zosimos's predecessors called Maria the Prophetess or Maria the Jewess. Today, when a pan of hot water is used to heat another vessel in the kitchen, the contrivance is still called a *bain–Marie*.

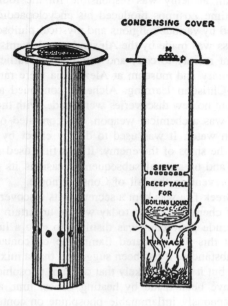

Figure 2.3 A reconstruction of a kerotakis

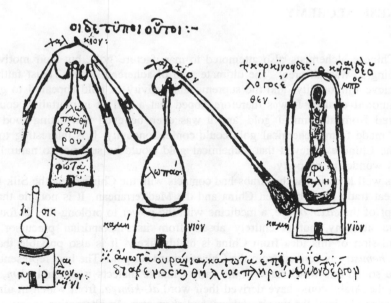

Figure 2.4 Greek alchemical apparatus

Early Alexandrian alchemy was responsible for the foundation of laboratory chemistry. By the time Zosimos produced his encyclopaedic work, alchemy was becoming dominated by various religious and mystical philosophies, and as a result little further progress was made by the Alexandrian alchemists. Later Alexandrian writers were content merely to copy and comment on earlier work.

In AD 390 the library and museum at Alexandria were ransacked by Christians suspicious of non–Christian learning. Alchemy continued to be practised in the Hellenistic world, but no new discoveries were made, with the important exception of *Greek fire*. This was a chemical weapon, and consisted of a liquid that caught fire on contact with water. It was used to deadly effect by the Byzantines, who sprayed it towards the ships of the enemy. It was first used in AD 665 against the fleet of the Arabs, and on several subsequent occasions its use was probably the decisive factor in preventing the fall of Constantinople.

The recipe for Greek fire was kept a secret, so its discovery had little impact on the development of chemistry. Even today we are uncertain of its composition. It seems likely that crude petroleum was distilled to give a liquid similar to petrol (gasoline), and that this was rendered flammable on contact with water by the addition of other substances. It has been suggested that a mixture of quicklime and sulphur was added, but it is more likely that calcium phosphide was used. Calcium phosphide could have been made by heating bones, lime and urine together. It produces the spontaneously inflammable phosphine on contact with water.

CHINESE ALCHEMY

The Chinese alchemists also attempted to manufacture gold, but their motivation for doing so was different. The ultimate goal of adherents to the Taoist faith was to achieve immortality, and the supreme unreactivity of gold appeared to give it an immortal quality. It was therefore hoped that a 'pill of immortality' could be prepared from alchemical gold, and it was even believed that eating food from plates made from alchemical gold could confer longevity. It is interesting to note that the Chinese believed that alchemical gold would be superior to natural gold in this wonderful property.

It is well known that the Arabs had contacts with the Chinese via the Silk Road, the great trade route between China and the Mediterranean. It is notable that the concept of the *elixir of life*, a medicine with the power to prolong life, is found in Islamic alchemy, but is entirely absent from its Alexandrian precursor. The transmission of this idea from China is highly likely. It is also possible that the word *chemistry* is derived from the Chinese word for gold. The best representation for the sound this word made in the early Chinese dialects is probably *kim*, from which the Arabs could have derived their word *al–kimiya*, from which in turn the Europeans obtained the words *alchemy* and *chemistry*. An alternative theory is that the Arabs derived *al–kimiya* from *Khem*, which means *Black Land*, and is the ancient name for Egypt.

ISLAMIC ALCHEMY

When the prophet Mohammed died in 632, most of the Arab tribes had been united under the new religion. There followed a period of rapid expansion for Islam, with the conquest of Syria in 640 and Egypt in 641. In 762 the Khalif al–Mansur founded a new capital at Baghdad for the Islamic world, which now stretched from Spain to Persia. With the encouragement of the Khalifs, Baghdad became a great centre of learning. Alexandrian alchemical manuscripts were translated into Arabic, and chemical phenomena excited the interest of men who were enthusiastic to develop the subject further.

The most important collection of Arabic alchemical texts was supposedly written by Jabir ibn Hayyan, but it has now been established that the writings, known collectively as the *Jabirian Corpus*, were the work of a Muslim sect known as the *Isma'iliya*. The corpus was completed by 987, but was probably compiled over a period of several generations. Jabir himself died about 813, but how much, if any, of the Jabirian Corpus is due to him personally is not known. The Jabirian works subsequently exerted great influence in Europe, where they were known in Latin translation and attributed to *Geber*.

The Jabirian theory of the formation of metals was clearly based on the views of Aristotle but included a significant new idea. Aristotle had considered metals to be formed by the combination of moist and dry exhalations, and in the Jabirian works these exhalations are identified with the vapours of mercury and sulphur. The cause of the different metals was the different quality of the sulphur from which they were formed. We must beware of identifying the sulphur of the Islamic alchemists with the pure material that we know by that name. The term *sulphur* probably embraced a whole range of sulphurs of varying purity and colour, and when used as a component of metals probably referred to a volatile combustible material to which no known substance corresponded exactly. Likewise mercury as we know it may only have been considered an approximation to the other volatile liquid component of metals. The sulphur–mercury theory is extremely important in the history of chemistry. The notion that metals contained a combustible principle persisted, and in European chemistry provided the inspiration for the phlogiston theory.

The Jabirian alchemists also believed that metals were ultimately composed of the four Aristotelian elements earth, water, air and fire, and in consequence possessed the qualities of coldness, hotness, dryness and moisture in varying proportions. They adopted a rational approach to the problem of transmutation. A base metal had to be treated with a medicine or *elixir* to adjust the qualities present to coincide with the proportions of gold. The existing proportions of the qualities in a base metal were estimated by means of rather obscure calculations.

Hand in hand with this was the idea that the qualities of heat, cold, moisture and dryness could each be separated in a pure form. The Jabirians attempted to do this in two stages. First they subjected various organic materials to dry distillation (Figure 2.5), which often resulted in the formation of a volatile combustible substance (air), a liquid (water), a combustible tarry material (fire) and a dry residue (ash). These elements were supposed to be composed of two qualities, and the Jabirians assumed that the predominant quality could be isolated by extended purification. Thus water, which is cold and moist with cold predominating, could be converted into pure cold by repeated distillation. Water had to be subjected initially to 70 distillations, and then to a further 700 in the presence of a drying agent. The resulting pure cold was supposed to be a brilliant white solid. Once he had obtained his pure elements, consisting of only one quality, the alchemist was supposed to mix them in the correct proportions to obtain an elixir that he could use to treat a base metal.

Figure 2.5 Islamic glass distillation apparatus, tenth to twelfth century

The Jabirian Corpus thus contained directions for transmutation which were logical but were extremely difficult to carry out. When an attempt at transmutation failed, the alchemist could blame his own experimental technique rather than the underlying theory. Apart from the Jabirian Corpus, the alchemy of Islam, like that of the Hellenistic world and of Europe, produced many writings of an obscure and mystical nature. However, the works of two authors are straightforward and clear from ambiguity and mysticism. These writers are al-Razi (864–925) and ibn-Sina (980–1037). They are usually known by their Latinised names of Rhases and Avicenna.

Rhases adopted the sulphur–mercury theory of metals, and also believed in the possibility of transmutation, but it is clear that he was a very knowledgeable practical chemist. He described the preparation of caustic alkalis by treating sodium or potassium carbonates (obtained by leaching ashes) with slaked lime. Our word *alkali* comes from the Arabic *al–Quili*, meaning calcined ashes. The caustic alkalis, along with acidic solutions such as vinegar, sour milk and lemon juice, were known as *sharp waters*, and were widely used as solvents. Rhases drew up a classification of substances which was more elaborate than those produced previously. He divided materials into animal, vegetable, mineral and derivative, and he further divided mineral bodies into six groups:

Spirits: volatile materials such as sulphur, mercury, sal ammoniac and arsenic.
Bodies: the metals (except mercury).
Stones: malachite, haematite, etc.
Vitriols: e.g. green vitriol (iron(II) sulphate).
Boraces: borax, soda.
Salts: common salt, potash, slaked lime, etc.

Chemists still classify materials into groups containing substances with similar properties as an aid to understanding them.

Avicenna is the last of the important writers of the Islamic period of alchemy. He is primarily remembered as a physician, and his medical text was regarded as highly authoritative for 600 years after his death. In his chemical writing he dismisses the possibility of a transmutation being achieved by the administration of an elixir to a base metal. If a transmutation is to be achieved at all, the metal must first be broken down into its constituent elements, and then an appropriate recombination attempted. Avicenna's views exerted an important influence on later European alchemy.

EARLY EUROPEAN ALCHEMY

In the twelfth century, European scholars became aware of the scientific and medical texts available in Arabic. The first translations of Arabic alchemical works into Latin were performed in Spain and Sicily, where Arabic and European civilisations came into contact. The first translator was probably the English cleric Robert of Chester, who was active around 1144. By the end of the twelfth century, several texts had been made available to European scholars.

In the thirteenth century, alchemical and scientific theories learned from the Arabs were incorporated into the encyclopaedias written by men such as Vincent of Beauvais (1190–c. 1264) and Albertus Magnus (1193–1280). These writers accepted the Aristotelian elements and the sulphur–mercury theory of the Arabs. They believed that transmutation was a possibility, but that it was very difficult to achieve. Albertus Magnus clearly doubted that a successful transmutation had ever been performed. These books acted as the starting point for many later European alchemists.

At about the time that the Arabic works were becoming known in Europe, important advances were being made in the art of distillation. Some time in the twelfth century, Europeans distilled wine in the presence of various moisture-absorbing salts and produced a distillate which was sufficiently concentrated in alcohol to be inflammable. In the thirteenth century the efficiency of condensation was improved by Thaddeus Alderotti (1223–1303) who introduced a water–cooled condenser. The coiled condensing tube was about a metre long and immersed in a vessel through which a constant flow of cold water was maintained (Figure 2.6). The inflammable alcoholic solution produced by these techniques was known as *aqua ardens*, and repeated distillation produced *aqua vitae*, or water of life. It was soon used medically in much the same way as brandy is today.

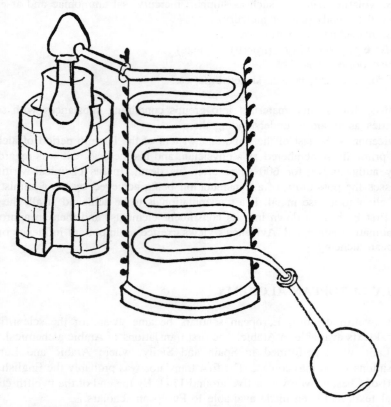

Figure 2.6 A distillation apparatus employing a water cooled condenser. Such apparatus was probably first used in the thirteenth century

One of the ideas that arose in the early phase of European alchemy was that all substances were composed not only of the four terrestrial elements but also of a fifth, which Aristotle had maintained was confined to the heavens. This fifth element was known as the *quintessence* of the substance in which it occurred. The marvellous properties of alcohol led to the belief that it was the quintessence of wine. Furthermore, alcohol was the first organic solvent known, and could extract essential oils from plants. These, with their distinctive odours, seemed to be the quintessences of the plants. Thus the quintessences of herbs and other plants could be extracted by solution or distillation with alcohol to form a liqueur. The earliest liqueurs were prepared for their medicinal properties; only later was their use appreciated in a purely social context.

Another important discovery of the early thirteenth century was that of the mineral acids. Vitriols were heated and for the first time the vapours were condensed, yielding concentrated sulphuric acid. This advance probably occurred not so much as a result of improved techniques of condensation but as a consequence of the ability to construct apparatus that would stand up to the corrosive product. Heating a vitriol with nitre (potassium nitrate) resulted in the production of nitric acid, which soon found a use in separating silver and copper from gold. The addition of sal ammoniac to nitric acid yielded *aqua regia*, which would dissolve gold itself.

Nitre was obtained from manure. The manure was formed into heaps, and nitre was produced by bacterial action. The nitre was then extracted by water and concentrated until common salt (the main impurity) crystallised out. The solution was then filtered, and crystals of reasonably pure nitre then formed. Nitre was also required for the manufacture of gunpowder, in which it is mixed with charcoal and sulphur. Gunpowder appears to have been a Chinese invention, for it is described in Chinese texts of the tenth century. Knowledge of gunpowder reached Europe in the thirteenth century.

We have seen how the idea of the elixir of life appeared in Islamic alchemy, probably by transmission from China. As well as prolonging human life, the elixir was deemed to have the power to cure 'sick' metals, making them more goldlike. European alchemists used the terms *elixir of life* and *philosopher's stone* almost interchangeably. The discovery of aqua vitae, with its power of prevention of decay as well as its medicinal properties, must have seemed to be a step nearer to the discovery of the true elixir of life.

Alchemical theory and practice changed little in the fourteenth and fifteenth centuries. The true alchemists, or *adepts*, were becoming increasingly concerned with their own spiritual perfection. They believed that the achievement of this end would be symbolised by the ability to make gold, and they called their endeavour the *Great Work*. At the other end of the scale were the *puffers*, who incessantly pursued the goal of fabulous riches, which would ensue once the technique of transmutation had been mastered.

TECHNICAL TREATISES

Although the fourteenth and fifteenth centuries saw only minor chemical progress, a technological advance occurred that was to have a dramatic impact on almost every aspect of human endeavour. This was the invention of printing with movable type, which took place about 1450. It is estimated that between 1450 and 1500 about 40,000 books were published, which is more than had appeared in the previous 1000 years.

Books became available on all manner of topics about which information had previously been hard to acquire. The technical aspects of chemistry formed the subject for several books that were published in the sixteenth century. In 1500 Hieronymus Brunschwygk published his *Little Book of Distillation*, which was followed by an enlarged version twelve years later. These books described the various techniques of distillation in detail, and were copiously illustrated. Other books on distillation soon followed.

Around 1500 there appeared two anonymous books, *Ein Nützlich Bergbüchlein* and the *Probierbüchlein*. The former is chiefly concerned with mineralogy, and the latter with the assaying of metals. The Probierbüchlein describes accurate quantitative methods using the balance. Over the next 50 years there appeared many books on mineralogy, mining and assaying. Three of the most famous were *De La Pirotechnia* of Vannuccio Biringuccio (1480–1539) published in 1540, *De Re Metallica* of Georg Bauer (1494–1555, also known as Agricola), which was published in 1556 (Figure 2.7), and the *Treatise on Ores and Assaying* of Lazarus Ercker (died 1593), which was published in 1574. Between them, these books give a complete picture of the chemical technology of the sixteenth century. They describe the preparation of sulphur, the manufacture of the mineral acids, the extraction of mercury, and the preparation of other materials that were used in metallurgical processes. With the aid of their clear instructions, anyone could repeat the work they described. They must have assisted considerably in the expansion of the infant chemical industry.

IATROCHEMISTRY

The early European alchemists had demonstrated that the products of a distillation might have medical applications, and early in the sixteenth century much greater attention was directed towards the medical applications of alchemy. This subdivision of alchemy is called *iatrochemistry*, and its leading practitioner was Paracelsus (1493–1541). He maintained that alchemy should include all processes by which a substance is changed and made fit for a new purpose. Although he believed in the possibility of transmutation, he widened the scope of alchemy to include all changes which we would now regard as chemical. In particular, he concentrated on the production of drugs and medicines in the laboratory, and advocated the use of a number of metallic salts in medical practice. He also believed that digestion was an alchemical process, which he thought was under the direction of a being called the *Archaeus*.

Paracelsus (Figure 2.8) is one of the most extraordinary and outrageous figures in the history of chemistry. His real name was Theophrastus Bombast von Hohenheim, and he arrogantly called himself Paracelsus to emphasise his imagined superiority to the Roman medical authority Celsus. Born in Switzerland, he studied alchemy, mining, magic and astrology before travelling widely in Europe. During his wanderings he gained the degree of Doctor of Medicine, and in 1527 he was appointed city physician at Basel. At his first public lecture he burnt the works of Avicenna and Galen, the recognised medical authorities of the day. He was forced to leave the town after about one year in office, and he then resumed his wanderings until his death. Paracelsus's writings are for the most part obscure, which is not surprising since they were mainly dictated while the author was drunk. It is, however, possible to extract his most important ideas, and there is no doubt that he exerted a considerable influence on chemistry.

Figure 2.7 Agricola's illustration of one of the methods of extracting mercury from its ore (cinnabar). The crushed ore was placed in the rounded pots. The mouths of the pots were filled with moss, and the pots were then inverted in the flat dishes. The joins were then luted (sealed with clay), and the vessels were placed in a hearth and surrounded with a mixture of crushed earth and charcoal. A fire was lit on top of the pots, and the liberated mercury percolated through the moss and collected in the flat dishes. Agricola was clearly aware of the poisonous nature of mercury vapour. He stressed the need for good-quality pots to be used, and stated that the loss of mercury from cracked pots was detectable by the sweet smell, and that this could loosen the teeth of the workers!

Paracelsus made an important contribution to chemical theory. He extended the sulphur–mercury theory of the Islamic alchemists by adding a third principle, namely salt. Thus, when wood burned, the combustible component was identified with sulphur, the volatile component with mercury and the ashes that remained with salt. The composition of all substances could be expressed in terms of these three principles, or *tria prima*. As in the previous theories, sulphur, mercury and salt were not equated with the common materials with those names, but rather with their essential qualities.

Figure 2.8 Paracelsus (1493–1541)

Paracelsus's works began to appear in print about 20 years after his death, and other authors soon produced iatrochemical texts. Probably the most famous such author was Basil Valentine, although it is now believed that no such person existed and that his books were probably the work of the publisher, Johann Thölde. The most famous book attributed to Valentine is the *Triumphal Chariot of Antimony*. This work gave clear directions for the preparation of a number of antimony salts and their use in medicine.

The significance of the iatrochemists is that they helped to direct both medicine and chemistry along new paths. Perhaps their most important legacy was that in the seventeenth century chemistry began to be taught in the medical faculties of many European universities, and as a result the subject started to acquire an academic respectability.

LATER EUROPEAN ALCHEMY

Alchemy reached its zenith in sixteenth and early seventeenth century Europe (Figure 2.9). The subject became increasingly mystical, but the appearance of printed books on alchemy acted as a spur for greater numbers to attempt to become adepts. The goal of alchemists at this time was summarised by Pierre–Jean Fabre in his book *Les Secrets Chymiques,* which was published in 1636:

'Alchemy is not merely an art or science to teach metallic transmutation, so much as a true and solid science that teaches how to know the centre of all things, which in the divine language is called the Spirit of Life.'

Figure 2.9 *The Alchemist's Experiment Takes Fire* by the seventeenth century Dutch artist Hendrick Heerschop. Accidents such as this must have been commonplace

Many alchemical texts of this period are copiously illustrated; indeed. some consist entirely of pictures. It is now realised that in some cases the words and pictures in these books, in spite of their apparent obscurity, refer to genuine chemical processes. An example is the 24th emblem from Michael Maier's *Atalanta fugiens,* which was published in 1618 (Figure 2.10). The description of the process depicted reads: 'The grey wolf devours the King after which it is burned on a pyre, consuming the wolf and restoring the King to life'. This is in fact a description of a method of purifying gold. The wolf is antimony sulphide, which would react with other metals present in the gold to yield their sulphides and free antimony. The sulphides would then be skimmed off the molten mixture. When the gold–antimony alloy is then heated, the antimony is vaporised, leaving pure gold.

Figure 2.10 An early seventeenth century pictorial representation of a process for purifying gold from Michael Maier's *Atalanta fugiens* of 1618

Partial interpretations are possible of other writings and illustrations that purport to describe the preparation of the philosopher's stone and transmutation. One of the most famous sequences is of 20 pictures from the *Philosophia Reformata* of John Daniel Mylius published in 1622 (Figure 2.11). Attempts to prepare the stone often started with gold and silver, which are represented by a king and queen, and also by the sun and moon. The king and queen are depicted as getting married, which presumably represents a fusion process. The king and queen die and the corpse (which becomes a hermaphrodite) decays. This presumably represents chemical attack on the gold–silver alloy by a reagent. Ultimately the regenerated king and queen produce a child, which is a symbol for the philosopher's stone.

A number of other images recur frequently in sequences of alchemical illustrations. Among these are the ascent and descent of birds (distillation and condensation?), a tree bearing fruit and dew descending from heaven. We may surmise that all these images represented chemical processes to the alchemists. Certainly anyone who thought he had interpreted an entire sequence of pictorial or written instructions and attempted to make gold would have met with disappointment. Not a few people brought themselves and their families to ruin as a result of their obsession with goldmaking. Brand's isolation of phosphorus in 1669 occurred when he used human urine in one of his attempts to prepare the philosopher's stone. The formation of phosphorus, with its inflammable nature and luminous emissions, must have given him reason to be optimistic.

Figure 2.11 The fifth picture from the sequence of twenty in the *Philosophia Reformata* of John Daniel Mylius (1622). The King and Queen (sun and moon, gold and silver) are married. The sun and moon are being attacked by black birds, and this represents their impending decay

FRAUDULENT ALCHEMISTS

Not surprisingly, alchemy provided a fertile ground for fraud. Stories abound of innocent individuals being duped by confidence tricksters posing as alchemists. One famous anecdote from the Islamic period relates how a dishonest Persian visitor to Damascus fooled the Sultan. The Persian prepared some pellets consisting of finely ground gold, charcoal, flour and glue. Adopting a disguise, he visited an apothecary and sold him the pellets, pretending they consisted of a rare and sought–after material called Tabarmaq of Khurasan. The Persian then donned expensive clothes and let it be known that he was an expert alchemist, and could make large quantities of gold. The Sultan heard of these claims and arranged for the Persian to demonstrate his skill. The Persian gave the Sultan's servants a list of his requirements, which included the mysterious Tabarmaq. After much searching, this was located at the apothecary's shop. The Sultan watched as the Persian ordered the ingredients to be strongly heated in a crucible. Everything in the mixture, with the exception of the gold, either vaporised or was oxidised to give volatile products, so when the crucible was removed from the furnace it contained only molten gold.

The Sultan rewarded the Persian and decided to try his own hand at goldmaking. All the ingredients were readily acquired except the Tabarmaq. The apothecary had sold his entire stock for the previous trial, and no one else in the city had heard of the mysterious substance. The Persian announced that he knew a cave in Khurasan where it was mined, and although he feigned reluctance he was eventually persuaded by the Sultan to go on an expedition to acquire a large quantity. The Persian duly departed, equipped with camels, a tent, carpets, silks and a large quantity of money. Not surprisingly, he was never seen again.

Many other ruses were used to create the illusion of transmutation. A trick of some dishonest European alchemists was performed with a nail, one half of which was made from gold and the other from iron. The gold portion was covered with black paint, and during the 'transmutation' the painted portion would be immersed in a solvent, which removed the paint and revealed the gold.

During the heyday of European alchemy some rulers supported alchemists while they pursued their researches. The most famous royal patron of alchemy was Emperor Rudolph II (died 1612), and in his capital city of Prague there is a 'Golden Lane', which was where his alchemists had their laboratories. No doubt Rudolph hoped to share in any successes achieved by the alchemists, but he also supported other scientists such as the astronomers Tycho Brahe and Johannes Kepler. Alchemists seeking patronage or rewards from potentates had to be careful not to promise too much. In 1597 Georg Honauer convinced Duke Frederick of Wurtemburg that he could turn base metals into gold. Honauer was provided with a large quantity of iron to transmute, and when he failed the Duke ordered the iron to be covered with gold leaf and made into a gallows. The dishonest alchemist was duly hanged from the gleaming structure.

Not surprisingly, the activities of the tricksters gave alchemy a bad name, and increasingly alchemists became the subject of ridicule. The English dramatist Ben Jonson wrote his play *The Alchemist* in 1610. The main characters are the dishonest alchemist Subtle and his assistant Face, who try to cheat Sir Epicure Mammon into investing money in their non–existent gold–making process. It is clear from the text that the audience was expected to be familiar with alchemical vocabulary, and that alchemists were commonly regarded as rogues.

In the eighteenth century various cranks and charlatans continued to proclaim that they had manufactured gold. One of the last was James Price, whose motive for making false claims seems to have been fame rather than fortune. In 1782 Price published a pamphlet detailing how he was able to transmute mercury into silver or gold using mysterious white and red powders. The news created a sensation in the scientific world as Price was known to be a chemist of some ability and had been elected a Fellow of the Royal Society. The Society's President insisted that Price should demonstrate his transmutations before an official committee of Fellows. Price agreed with great reluctance. When the appointed day came, the witnesses were shown into Price's laboratory. Price left the room for a moment and quickly drank a solution of prussic acid. Returning to the laboratory, he died in front of the committee. Perhaps the ultimate irony of the saga of James Price is that the modern atomic physicist, by placing mercury in the core of a nuclear reactor, has succeeded where Price and all other alchemists failed. Unfortunately, the isotope of gold produced (^{200}Au) is unstable, with a half–life of only 48 minutes, but at least the 'Great Work' is now technically possible.

3

From Alchemy to Chemistry

The seventeenth century was the period in which modern science began to emerge, and its evolution from its mediaeval precursor is usually termed the *Scientific Revolution*. Like most revolutions in human history, it was not a sudden or simple process. If we are to attempt to understand the activities and beliefs of the alchemists and chemists of this period, we must view their work against the background of this revolution. By the end of the seventeenth century the new science of chemistry was providing fresh explanations for the properties of materials and their transformations, and the alchemist's dream of making gold was beginning to fade.

THREE TRADITIONS

The Scientific Revolution can be regarded as a battle between three different ways of looking at the natural world. These can be termed the Aristotelian, the magical and the mechanical. All three traditions had their origins in either the Greek or the Hellenistic periods. It was the ultimate triumph of the mechanical philosophy which was to result in the birth of modern science.

We have seen how Arabic alchemical works began to appear in translation in Europe in the twelfth century. At the same time, translations of other works on the science and philosophy of the ancient world were made. This new learning, principally of Aristotle in the physical sciences, Galen (AD 129–199) in medicine and Ptolemy (AD 90–170) in astronomy, initially presented a challenge to the Church, but it was gradually assimilated into Christian orthodoxy. The most prominent cleric in this movement was Thomas Aquinas (1226–1274), and by 1500 the process was complete. According to the Aristotelian view, change was continually occurring in the sub–lunar world, and was controlled by final causes. The heavens were constant and unchanging, and the earth was situated at the centre of the universe with the moon and planets rotating around it, and the stars held on a fixed sphere. Substances were supposed to possess properties of two types: elementary and hidden (or *occult*). Elementary properties were those such as colour, density, taste, etc., and were supposed to depend upon the nature of the substance. Occult properties were those which were not understood, and were held to be incapable of explanation. They were supposed to be the arbitrary decision of the Creator, and could not be influenced in any way. Examples were the medicinal properties of substances and magnetism. Chemical change was explained in terms of the four–element theory, and every chemical reaction was viewed as a kind of transmutation.

Some writings of the ancients only became known in Europe after the fall of Constantinople in 1453. Among these were a collection of works originally attributed to Hermes Trismegistos, which were thought to be an exposition of Egyptian thought from the time of Moses. Hermes Trismegistos was supposed to have received divine revelations about the nature of the physical world. About 100 years after the Hermetic writings became known in Europe they were shown to date from about the third century AD, and to be the work of the neo–Platonists under whose influence Alexandrian alchemy had become more mystical in character (Chapter 2).

The Hermetic writings formed the basis of the magical tradition in European science. The revival of these mystical and magical ideas received some sympathy from Christian theologians, as miracles were much easier to explain than under the Aristotelian system. The renewed interest in neo–Platonism exerted an influence on many of the later European alchemists and iatrochemists.

In the period leading up to the Scientific Revolution, a third way of looking at the universe began to emerge. This was the mechanical philosophy, which became popular partly as a result of the appearance in the mid–sixteenth century of a printed edition of the works of Archimedes (287–212 BC), who had not only been a great mathematician, but had also been fascinated by mechanical analogies. As a result of the revival of interest in Archimedes's work, some philosophers began to view the universe as a giant mechanism governed by unchanging scientific laws, which were capable of mathematical expression. In this tradition, God acquired the attributes of an engineer.

Prominent among early adherents to the mechanical philosophy was Galileo Galilei (1564–1642). For his mechanistic explanations, Galileo drew upon the atomistic views of Democritos (Chapter 1). Another influential mechanist who attempted to explain the properties of matter in atomic terms was Pierre Gassendi (1592–1655), who made a careful study of the works of Epicurus (Chapter 1). Not all mechanists accepted the atomic hypothesis. René Descartes (1596–1650), in his *Principia philosophiae* of 1644, maintained that matter was infinitely divisible. Space was filled with a very fine, subtle, continuous matter, which by its whirling motion carried planets around the sun. Robert Boyle (1627–1691) was to come down on the side of the atomic theory of matter in what he called his *corpuscular philosophy*. However, explanations of certain physical phenomena (e.g. the propagation of light) were to be advanced in terms of a subtle, continuous matter, or *aether*, right up to the beginning of the twentieth century.

NEW IDEAS

The seventeenth century saw the rise of experimental science. The notion of making observations and performing experiments was not, of course, a new one in the seventeenth century. However, the experimental approach was highlighted by Sir Francis Bacon (1561–1626), who emphasised that experiments should be planned and the results carefully recorded so that they could be repeated and verified. As a result of experimental work, generalisations could be made and theories proposed which would lead to further suggestions for experiments. Bacon also proposed the formation of societies where scientists could meet to report their observations and discuss their theories.

The sixteenth and seventeenth centuries saw many new ideas emerge which were in direct conflict with ancient dogma. The challenge to Galen came in 1543 when Andreas Vesalius (1514–1564) published his book *The Fabric of the Human Body*. As a result of his own careful dissections of human corpses (a practice outlawed in Galen's time), Vesalius was able to show that Galen had been wrong on approximately 200 points. Another attack on Galen came in 1628 when William Harvey (1578–1657) discovered the circulation of the blood. Galen had supposed the arteries and veins to be two almost separate systems, with some degree of mingling of the blood through pores in the septum dividing the heart. Harvey realised that the valves in the veins must mean that the blood flows in one direction only, and by measuring how much blood was ejected from the heart in each pulse he came to the conclusion that the blood must move 'as it were, in a circule'.

The year 1543 also saw the appearance of another book that not only challenged the ancient authorities, but was also to have a profound impact on the way people perceived their place in the universe. The book was *De revolutionibus orbium coelestium* by Nicolaus Copernicus (1473–1543). On the basis that predictions of the future positions of the planets made using Ptolemy's geocentric theory were inaccurate, Copernicus proposed that the sun was at the centre of the universe, and the planets, including the earth, revolved around it. Copernicus still believed in the sphere of the stars, but man's unique position at the centre of all things was gone.

Copernicus was undoubtedly influenced by the revival of neo–Platonism, and in his writings he refers several times to Hermes Trismegistos. Subsequent revolutionary figures in astronomy were also in the magical tradition. While Copernicus made few observations of his own, Tycho Brahe (1546–1601) amassed a vast collection of careful observations with the aim of providing accurate horoscopes for his royal patrons. Brahe did not publish his observations, but used his private collection of data to make for himself a leading position in the world of astrology. However, on his death–bed he gave his data to his assistant, Johannes Kepler (1571–1630).

Before joining Brahe, Kepler had published a book in which he proposed that the structure of the planetary system was based on the five regular polyhedra, an idea clearly derived from the Pythagoreans. Using Brahe's observations Kepler formulated his first two laws of planetary motion. These state that the planets move in elliptical orbits with the sun at one focus, and that a line from the sun to a planet sweeps out equal areas in equal times. These ideas were published in *Astronomia Nova* in 1609. Kepler's third major work, *Harmonice Mundi*, was published in 1619. As well as containing his third law (the square of the planetary period of revolution is proportional to the cube of the mean distance of the planet from the sun), the book contained a great deal of mystical speculation. His most famous idea was that each planet had a musical scale associated with it, the pitch of the lowest and highest notes being related to the minimum and maximum velocity of the planet in the orbit.

It was Galileo who in 1609 first viewed the heavens through the newly discovered telescope and obtained evidence supporting the heliocentric view of the solar system. It had been argued that, if the heliocentric view were correct, it was unlikely that the earth would be unique in having a moon rotating around it. On the geocentric view, the sun, moon and planets were all satellites of the earth. Galileo was the first to see the moons of Jupiter and observe their rotation around the planet, and in consequence he came into conflict with the Roman Catholic church.

The most famous scientist of the seventeenth century was Sir Isaac Newton (1642–1727). His concept of gravitation applying to all the objects in the universe can easily be seen as a triumph for seventeenth century mechanism, but it is incorrect to suppose that Newton had a purely mechanistic view of nature. He was, in fact, strongly influenced by the magical tradition, and this is evidenced by the enormous amount of time he spent on alchemical research. At intervals over a 25–year period, Newton conducted experiments in his alchemical laboratory at Trinity College in Cambridge and left a mass of manuscripts describing his experiments. He copied portions of earlier alchemical manuscripts and made great efforts to decode them. Newton's sole published work on chemistry was *De natura acidorum* (On the nature of acids, 1710) in which he attempted to account for acidity in terms of the attractive forces between particles.

It can therefore be seen that the discoveries of the sixteenth and seventeenth centuries were being made against the background of a conflict between the Aristotelian, the magical and the mechanistic traditions. During this period, most alchemists and iatrochemists were influenced by the magical neo–Platonic doctrines. Until the middle of the fifteenth century the speculations of the alchemists had usually been couched in Aristotelian terms, and the concept of the four elements had made transmutation a theoretical possibility. The dissemination of writings attributed to Hermes Trismegistos and the consequent revival of neo–Platonism were influential in making alchemy more mystical.

THE LATER IATROCHEMISTS

The alchemists of this period who made useful contributions to the development of chemistry were not totally distracted by the attempt to make gold artificially. Most of these workers were iatrochemists, and Paracelsus has already been mentioned (Chapter 2). He too was influenced by the renewed interest in neo–Platonism, and he applied the neo–Platonic doctrine of the microcosm and macrocosm to medicine. He saw the human body as a microcosm of all that existed in the universe (the macrocosm), and thus the organs of the body were the equivalents of the stars. Such mystical ideas resulted in Paracelsus concentrating on particular areas of the body and suggesting specific remedies for individual diseases, rather than advocating general measures such as blood–letting.

Paracelsus initiated a debate on the place of chemistry in medicine. The more traditional members of the medical profession clung to the ideas of Galen, but from the middle of the sixteenth century onwards an increasing number of more progressive physicians produced iatrochemical texts advocating the use of chemically prepared remedies. Typical of these iatrochemical writers was Joseph Duchesne (*c.* 1544–1609), also known as Quercetanus, a physician who advocated the medical use of substances such as antimony sulphide, urea and calomel, and gave directions for their preparation.

In 1597 there appeared the first true textbook of chemistry, the *Alchemia* of Andreus Libavius (1540–1616). Libavius was a teacher and physician interested in chemistry, and in preparing his book he drew upon the writings of alchemists, iatrochemists, pharmacists and metallurgists. He defined alchemy as 'the art of extracting perfect magisteries and pure essences from mixed bodies'. By a *magistery* Libavius meant an extract or concentrate displaying the essential properties of the starting material. He added that alchemy is 'valuable in medicine, in metallurgy and in daily life'. The book was divided into two parts. The first, *Encheria*, dealt with chemical procedures, and the second, *Chymia*, was concerned with the preparation of substances. Like his contemporaries, Libavius accepted the possibility of goldmaking, and he regarded many chemical changes as a kind of transmutation. But his book stressed the practical rather than the theoretical side of chemistry, and was important in assisting the emergence of chemistry as an independent science. His descriptions of practical recipes are clear, but he became obscure when discussing the transmutation of metals, stating that such secrets were possessed by only a few. Libavius drew up detailed plans for a purpose–built laboratory complex, although he did not construct such a building himself.

Another book which was influential in the seventeenth century was the *Tyrocinium chymicum*, or *Chemical Beginner* (1610), of Jean Beguin (*c.* 1550–*c.* 1620). Beguin set up a laboratory in Paris where he gave public lectures on the preparation of the new chemical remedies. His book was based on his lectures, and was more concerned with chemical operations than with theory. Beguin's book was the first of many French chemical textbooks which appeared in the seventeenth century.

One of the most important figures in practical chemistry in the seventeenth century was Johann Rudolph Glauber (1604–1670). Unlike many of his contemporaries, he was not medically qualified, and he learned his practical chemistry by visiting a number of laboratories in different parts of Europe. He devised new types of furnace, which enabled higher temperatures to be achieved. He is particularly remembered for his preparations of the mineral acids and their salts. He realised that oil of vitriol (sulphuric acid) prepared by burning sulphur under a bell jar was the same as that obtained by heating green vitriol (iron(II) sulphate). He obtained hydrochloric and nitric acids by the action of oil of vitriol on chlorides and nitrates. Glauber made some of his chemicals on a large scale, and he made his living by selling them. He can be regarded as an early industrial chemist. One of his most famous products was *sal mirabile*, hydrated sodium sulphate, for which he claimed marvellous medicinal properties. This compound is still known as *Glauber's salt*.

Perhaps Paracelsus's most important successor was Johann Baptista van Helmont (1577–1644). His work was important for its quantitative character, and he made extensive use of the balance. He realised that when a metal is dissolved in an acid, for example when silver is dissolved in aqua fortis (nitric acid), the silver is concealed in solution and can subsequently be recovered. When one metal precipitates another from solution, he realised that there is no transmutation as Paracelsus had thought.

> Van Helmont (Figure 3.1) was born in Brussels. He qualified as a physician but practised little; he was apparently sufficiently wealthy to devote most of his time to chemical research. He called himself a *philosopher by fire*. He made a careful study of the writings of Paracelsus, and while he was undoubtedly influenced by Paracelsus's work, he found it to contain many errors. Shortly before his death, he requested that his son should publish his collected papers, and the first edition appeared in 1648, entitled *Ortus medicinae*. Van Helmont's writings were influential on later workers, especially Robert Boyle. Van Helmont's contributions to chemistry were of the first importance, but nevertheless his writings contain an account of a transmutation of mercury into gold by a grain of the Philosopher's Stone given to him by a stranger. Van Helmont personifies the transition from alchemy to chemistry.

Figure 3.1 Johann Baptista van Helmont (1577–1644)

Van Helmont rejected earlier ideas on the nature of matter, such as the theories of the four elements and three principles. He believed that the two elements were air and water, but that air was incapable of being changed into any other material. This left water as the fundamental constituent of all other substances, and he justified this conclusion by means of a famous and brilliantly conceived experiment. He planted a willow sapling weighing five pounds in a pot containing 200 pounds of dried soil. For a period of five years he added only rain water or distilled water, and at the end of that time he recovered the willow and found that it weighed 169 pounds. He dried all the soil once more, and discovered that it now weighed only two ounces less than at the outset. He therefore concluded that the increase in weight was due to the water, which had been converted into the wood of the tree.

Van Helmont's conclusion, was, of course, mistaken because he was unaware of the uptake of carbon dioxide by the plant. There is an undoubted irony in this, because it was he who coined the term *gas*, and recognised gases as a new class of substance. He found that when 62 pounds of charcoal were heated, only one pound of ashes remained, and he deduced that 61 pounds of *gas sylvester* (carbon dioxide) escaped. Sometimes when gas was generated in a sealed vessel, for example when nitric acid came into contact with sal ammoniac, the liberated gas burst the vessel. Van Helmont used the terms *wild spirit* or *untamable gas* for such substances. He probably derived the word *gas* from the Greek word *chaos*. He clearly thought that it was impossible to contain these wild substances, but recognised that there were different types of gas with different properties. Among the other gases mentioned by Van Helmont were a red gas (nitric oxide) formed when aqua fortis (nitric acid) was reacted with silver, and *gas pingue*, an inflammable gas formed by the dry distillation of organic matter. The principal constituents of gas pingue would have been hydrogen, methane and carbon monoxide.

Van Helmont followed Paracelsus in recommending many mineral remedies for ailments. Like Paracelsus, he attempted to explain bodily processes such as digestion in chemical terms, and he modified Paracelsus's theory of the Archaeus (Chapter 2). He tried to analyse urinary calculi (stones) and to suggest the cause of their deposition in the body.

THE MECHANICAL PHILOSOPHY IN CHEMISTRY

The man who more than any other attempted to produce mechanistic interpretations of chemical phenomena was Robert Boyle (1627–1691). He backed up his ideas with ruthless logic and repeated reference to experiment. In his book entitled *The Sceptical Chymist*, published in 1661, he poured scorn on the idea that all compounds were composed of the same three principles or the same four elements.

For many years it had been believed that fire would break down materials into their elements, but Boyle pointed out that the number of products obtained varied with the method by which the fire was applied. Thus, when wood was heated in a retort, it yielded oil, spirit, vinegar, water and charcoal, but when heated in the open, it gave only ashes and soot.

Boyle (Figure 3.2) was born in Ireland, the fourteenth child of the first Earl of Cork. Between the ages of twelve and seventeen he lived in Geneva under the care of a tutor. During this period he made a trip to Italy, and while in Florence he studied the work of the recently deceased Galileo, and this led him to reject Aristotelianism. In 1644 Boyle returned to England and became an enthusiastic experimenter. He soon became sufficiently well known in scientific circles to make the acquaintance of a group which met to discuss the new mechanistic and anti–Aristotelian experimental science. This group called themselves *The Invisible College* and by the mid–1650s many were resident in Oxford. Boyle moved to Oxford in 1654, and the members of The Invisible College frequently held their discussions in Boyle's lodgings. The Invisible College was the forerunner of the Royal Society, which received its Charter in 1662 with Boyle as one of the members of its original Council. The Society provided a forum for scientists to meet and discuss their views, and it also published a journal (*Philosophical Transactions*). It thereby provided a considerable stimulus to scientific research of all kinds. In France, the Paris Academy of Sciences performed a similar role from 1666. Boyle moved to London in 1669, where he remained for the rest of his life.

Boyle believed that all matter was made up of the same kind of ultimate particles, and that the elements were composed of different groupings or 'primitive coalitions' of these particles. He gave no examples of elements, but believed that their number might be much larger than the three or four assumed by the alchemists. The properties of the elements were explained by the size, shape and motion of the groups of particles of which they were composed, and chemical reactions could be explained by the rearrangement of these units.

Boyle's destructive criticism of the old ideas had little immediate effect on contemporary chemical thought, as he was unable to propose a list of elements to replace the three principles or the four Aristotelian elements. His importance lies in the fact that he helped to liberate chemistry from the old modes of thought. He started to shift the emphasis of theoretical speculation away from *why* a chemical reaction occurred (which the Aristotelians did in terms of final causes) to *how* a reaction occurred (which Boyle attempted to do in terms of particles).

It will be recalled that the Aristotelians had argued that a vacuum was an impossibility, but experiments in the seventeenth century showed that a vacuum could indeed be created. It was Evangelista Torricelli (1608–1647) who suggested the experiment of filling a long glass tube with mercury and then inverting it in a trough of mercury. The mercury fell till the column was about 30 inches high, suggesting that the space above the mercury was a vacuum. The first air pump was devised by Otto von Guericke (1602–1686), who showed around 1654 that in an exhausted vessel the sound of a bell was muffled and that a flame was extinguished.

While at Oxford, Boyle employed as his assistant Robert Hooke (1635–1703), who constructed a considerably improved version of the air pump. Boyle and Hooke conducted many experiments on the properties of a vacuum, and also of the effect on the volume of varying the pressure of a certain quantity of air. The latter experiments led to the well known *Boyle's law*.

Using the air pump, Boyle conducted experiments on combustion. His method was to place a red–hot iron plate in a vessel, which he immediately evacuated, and then to arrange for a piece of combustible material to fall on the plate. Normally inflammable materials such as sulphur failed to catch fire in the vacuum, but did so when air was admitted. He made the important observation that gunpowder did ignite in a vacuum, and concluded that the nitre (potassium nitrate) in the gunpowder on heating gives 'agitated vapours which emulate air'.

Figure 3.2 Robert Boyle (1627–1691), with an air pump in the background

Boyle conducted experiments in which he calcined metals by heating them strongly in sealed glass retorts. When a retort was opened, he found that the calx (oxide) that had been formed weighed more than the original metal. He concluded that fire particles, which he thought had weight, had travelled through the glass and combined with the metal. Lavoisier repeated these experiments, but weighed the retort before opening (Chapter 5).

Robert Hooke (1635–1703) was originally intended for a career in the church, but his health was judged to be too poor and he turned to science. As a young man he worked as Boyle's assistant in Oxford. He then moved to London and was one of the original Fellows of the Royal Society. He was employed as *Curator of Experiments* to the Society, which involved preparing the experiments to be performed before the Fellows at the meetings. He was one of the first people to make a living from scientific employment. He was careless about his personal appearance; Samuel Pepys remarked in his *Diary* that Hooke 'is the most, and promises the least, of any man in the world that ever I saw'. Hooke claimed to have anticipated Newton in formulating the law of gravitation, and a bitter dispute broke out between them on this issue. He discovered the law named after him, which states that the deformation of a material is proportional to the force applied to it. He redesigned the microscope, and with its aid studied many minute objects and organisms, publishing his findings in his book *Micrographia* in 1665. He was one of the most versatile scientists of the seventeenth century, and, like Boyle, he was an enthusiastic supporter of the mechanical philosophy. No portrait of him survives.

Boyle produced hydrogen by dissolving iron in dilute sulphuric or hydrochloric acid. He collected the gas in an inverted glass vessel and noted its inflammability. He made important contributions to qualitative analysis. In his *Experimental History of Colours* (1665) he described many colour reagents and acid–alkali indicators.

Boyle was not alone in attempting to apply the atomic hypothesis to chemistry. Nicolas Lemery (1645–1715) suggested that the properties of substances could be explained in terms of the shapes of their atoms. The sharp taste of acids was due to their atoms being pointed and able to prick the tongue. Metals dissolved in acids because the points of the acid particles were able to break up the aggregation of metal particles. Lemery described his theories in his book *Cours de Chymie* published in 1675 (Figure 3.3). Lemery's book was also a comprehensive treatise on the practical chemical knowledge of the time. It ran to many editions and was very influential.

It was Hooke who put forward the first coherent theory of the phenomenon of combustion in the *Micrographia*. Hooke believed that combustible materials contained the sulphur principle and that air acted as a solvent for sulphurous bodies. He thought that during combustion a proportion of the inflammable material was 'dissolved and turned into the air, and made to fly up and down with it'. The act of solution produced heat and fire and was caused by a substance in the air, and a similar or identical substance was combined in nitre.

Further work on combustion was performed by another Englishman, John Mayow (1641–1679), who was a medical practitioner in Bath. Mayow recognised the similarity between combustion, respiration and the calcination of metals. In a book published in 1674, he put forward a theory of combustion well supported by experimental evidence. He found that only a portion of the air was involved in combustion by burning a candle in a limited quantity of air until it was extinguished, and then showing that a normally combustible material such as sulphur or camphor could not be ignited by means of a burning lens (Figure 3.4). Mayow stated that air and nitre contain *nitro–aerial spirit*, which supports combustion and respiration, and another constituent, which is inert. The sulphurous particles of the combustible material and the nitro–aerial particles of the air emit the heat and fire of combustion when they collide.

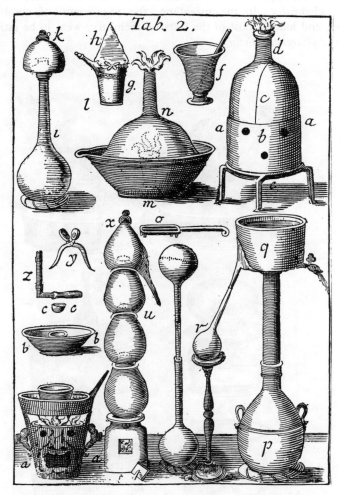

Figure 3.3 Chemical apparatus illustrated in Lemery's *Cours de Chymie* of 1675. The distillation apparatus at the bottom right is equipped with a *Moor's head*. This surrounds the alembic (still head) and is filled with cold water to bring about more efficient condensation

Mayow recognised that blood is heated as a result of respiration, not cooled as had been previously thought. Like Boyle and others before him, Mayow found that a metal increased in weight when it was calcined. However, while Boyle believed that this was due to the combination of the metal with fire particles, Mayow held that the metal was combining with nitro–aerial particles from the air.

It would be wrong for us to conclude that Mayow was on the threshold of discovering the oxygen theory of combustion when he died at the early age of 38. He did not heat nitre strongly, which would have generated oxygen, but even had he done so, there was as yet no apparatus for the manipulation of gases. Such experimental techniques were to be evolved in the next century in the age of pneumatic chemistry. In the meantime, the phenomena of combustion and calcination, and indeed most of the facts of chemistry, were to be rationalised in terms of the *phlogiston theory*.

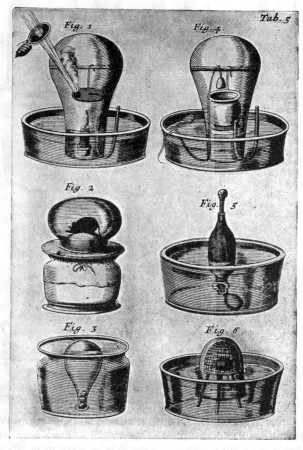

Figure 3.4 Some of the apparatus used by John Mayow. His experiment on combustion is illustrated at the top left. The apparatus shown at the bottom left is similar to that used by Boyle in his preparation of hydrogen

4

Phlogistic and Pneumatic Chemistry

The inadequacies of the old theories of the four elements and the three principles had been highlighted by Robert Boyle (Chapter 3). These theories had attempted to explain the composition of matter, but they could only account for chemical change in the vaguest terms. Boyle's corpuscular theory explained chemical change in terms of a rearrangement of particles, but it was of no use in systematising the chemistry of the day. Many workers had their own versions of the older theories, and it was the ideas of Johann Joachim Becher (1635–1682) that formed the basis for the formulation of the first comprehensive theory of chemistry.

THE PHLOGISTON THEORY

It had long been regarded as self–evident that combustion represented the decomposition of a material into simpler substances. In 1669 Becher combined this idea with his suggestion that all bodies were composed of air, water and three earths, which he called *terra pinguis, terra mercurialis* and *terra lapidea* (fatty earth, mercurial earth and stony earth). Becher proposed that terra pinguis escaped when combustion occurred.

Becher's theory might have exerted no more influence than the many similar ideas being expressed at the time had it not been taken up and extended by Georg Ernst Stahl (1660–1734). In 1703 Stahl replaced Becher's terra pinguis with *phlogiston*. The word was derived from the Greek word *phlogistos*, meaning burnt. Phlogiston was supposed to be a very subtle material, which could only be detected when it left a material containing it, and under such circumstances it appeared as fire, heat and light. Combustion was therefore the loss of phlogiston, and any remaining residue or ash was composed of the original material deprived of its phlogiston. Since nearly all metals could be converted to an ash–like material (calx) on strong heating, Stahl proposed that the calcination (oxidation) of a metal likewise occurred with the loss of phlogiston. Therefore, a metal was composed of its calx combined with phlogiston.

A large number of chemical facts could be explained by the phlogiston theory. Since charcoal burns in air leaving only a small quantity of ash, it was clear that charcoal was rich in phlogiston. The formation of a metal when its calx was heated

47

with charcoal could be explained by the phlogiston leaving the charcoal and uniting with the calx. It had been known for a long time that an inflammable substance would cease burning if the supply of air was limited, and Boyle had demonstrated that combustion would not occur in a vacuum. The phlogiston theory assigned to air the ability to absorb phlogiston, and, when the air was saturated with phlogiston, combustion ceased. Combustion was completely impossible in a vacuum with no air to absorb the phlogiston.

The phlogiston theory was the first unifying theory of chemistry, and it proved possible to use it to rationalise many new discoveries. One fact which could not be explained satisfactorily by the phlogiston theory was the well known increase in weight of metals on calcination. Many chemists were not concerned by this, because in the first half of the eighteenth century few were thinking in quantitative terms. It was enough for most that the theory gave a good qualitative explanation of chemical change and provided a unifying theory for many disparate chemical facts where none had existed previously. Towards the end of the century, some took refuge in the postulate that phlogiston had negative weight, and therefore its expulsion from a metal on heating would result in the solid getting heavier.

Not all chemists of the first half of the eighteenth century were converted to the phlogiston theory. Foremost among those who remained unconvinced was Hermann Boerhaave (1668–1738). Boerhaave was renowned as a teacher, and simultaneously held chairs in medicine, botany and chemistry at the University of Leyden. He lectured in all three fields, and also wrote excellent textbooks in each. It was his textbook of chemistry, the *Elementa chemiae* of 1732, which was to become the most influential. Boerhaave was only goaded into writing this book when some of his students published their own version of his lectures. Their book was an immediate success, but Boerhaave disowned it, pointing out that there were errors on every page. The *Elementa chemiae* appeared in numerous editions throughout the eighteenth century and was widely translated.

Boerhaave adopted a view of the nature of metals diametrically opposed to that of the phlogistonists. He rejected the idea that metals were compound bodies because, in spite of repeated attempts he was unable to decompose mercury and lead into simpler substances. He did not produce an alternative explanation of calcination to that of the phlogistonists, but remarked: 'We must check the Forwardness of the Imagination by the Weight of Experiments'. While the phlogistonists were speculative in their outlook, Boerhaave was practical and empirical. It was for this reason that Boerhaave's book remained highly influential for many years after his death, in spite of the fact that the phlogiston theory became almost universally accepted from the middle of the eighteenth century until Lavoisier's theory began to make converts.

Chemistry had won a place in the medical curriculum as a consequence of the views of the iatrochemists. Boerhaave sought to raise chemistry to an independent academic discipline. He drew students from all over Europe, many of whom subsequently held posts in the medical faculties of universities in their own countries and encouraged the teaching of chemistry. Although Boerhaave's own chemical discoveries were few, his influence on the development of the subject was immense.

TABLES OF AFFINITY

As the mass of chemical facts grew, attempts were made at systématisation by means of tables of affinity. These summarised the state of knowledge on reactions which we would term displacements. The first such table was produced in 1718 by Etienne–François Geoffroy (1672–1731). The table was arranged in vertical columns, and a substance at the head of a column would combine with all those beneath it, but the strength of the union decreased as the column was descended (Figure 4.1). Hence a substance would displace another beneath it when the latter

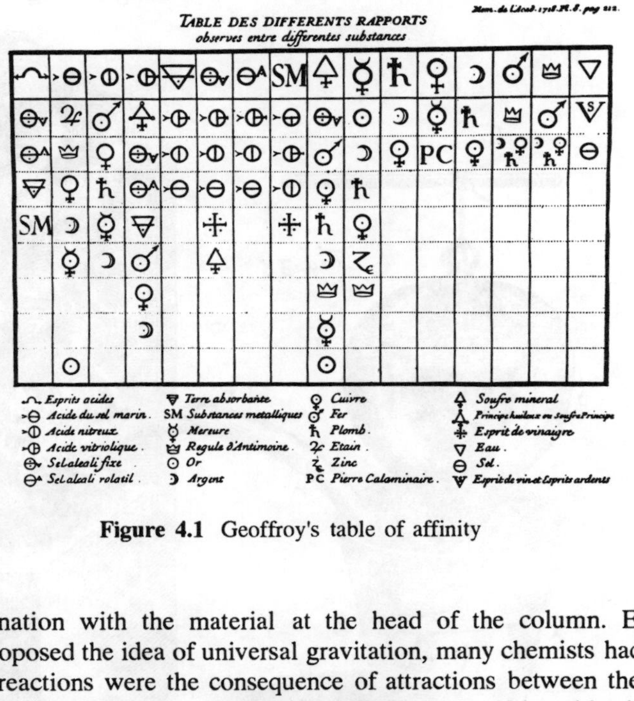

Figure 4.1 Geoffroy's table of affinity

was in combination with the material at the head of the column. Ever since Newton had proposed the idea of universal gravitation, many chemists had assumed that chemical reactions were the consequence of attractions between the particles of the reagents, and it was hoped that tables of affinity would enable the relative strengths of these attractive forces to be assessed.

The most comprehensive work on affinity was published in 1775 by Torbern Bergman (1735–1784). He produced two sets of tables, one for reactions at low temperatures with reagents in solution, and the other for reactions at high temperature when the reagents were in the dry fused state. He distinguished between *single elective attractions*, which were displacement reactions, and *double elective attractions*, by which he meant double decompositions. Bergman is perhaps most famous for the important contributions he made to qualitative and quantitative analysis (Chapter 14).

PNEUMATIC CHEMISTRY

Van Helmont had believed that a gas could not be confined in a container, but
Robert Boyle had collected factitious air (hydrogen) in the apparatus illustrated by
Mayow (Chapter 3). An advance in experimental technique was made by Stephen
Hales (1677–1761), who was a priest at Teddington, near London. His scientific
interests spanned botany and physiology as well as chemistry. The study of gases,
which Hales's work did so much to encourage, was given the name *pneumatic
chemistry*. Hales, like Boyle and others in Britain at this time, referred to gases as
airs, and he performed a series of quantitative experiments to ascertain how much
air could be expelled from various substances by strong heating.

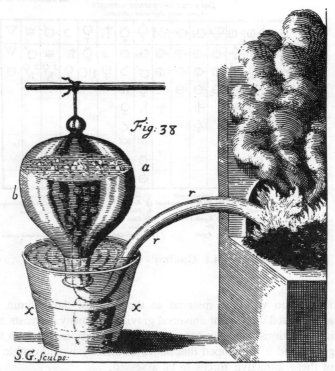

Figure 4.2 Hales's apparatus for the collection of the gas expelled from a substance by the
action of heat

Hales's apparatus was described in his book *Vegetable Staticks* of 1727, and
consisted of a bent gun barrel closed at one end. The substance to be heated was
placed at the closed end of the barrel, which conducted the gas evolved to an
inverted glass vessel full of water suspended in a tub which also contained water
(Figure 4.2). Hales thereby devised a way of generating and collecting gases in
separate vessels. Using this technique he collected the gases evolved when a wide
range of substances were heated, and he also obtained the gas from reactions such
as the action of dilute acids on iron filings. In each case he measured the volume

of 'air' produced from a weighed quantity of solid, but he made little attempt to investigate the properties of the airs he had isolated. He regarded all airs as consisting of atmospheric air with extra particles added; thus inflammable air consisted of common air with some additional inflammable particles.

Black (Figure 4.3) studied medicine at the Universities of Glasgow and Edinburgh, and in 1754 presented his dissertation for the degree of MD at Edinburgh. This contained his famous work on *Magnesia Alba* and was written in Latin. The English version was published in 1756, and had an enormous influence on the development of chemistry over the next 50 years. Black was appointed lecturer in chemistry at Glasgow in 1756, combining this appointment firstly with the professorship of anatomy and botany and subsequently with the professorship of medicine. In 1766 he became professor of chemistry and physic (medicine) at Edinburgh, where he remained until his death. His lecture courses in chemistry enjoyed a very high reputation and attracted audiences of several hundred students. He made some significant discoveries, which he did not publish but included in his lectures, the most important of which were connected with heat. He distinguished between heat and temperature and developed the concepts of latent and specific heat. While at Glasgow he became friendly with James Watt (1736–1819), who at the time was mathematical instrument maker to the university. Watt, who had not had a university education, learnt a great deal from Black, and the two conducted a lifelong correspondence on scientific matters. Watt's improvement to the steam engine came as a result of considering the heat losses that arose from alternate heating and cooling of the cylinder, and his discussions with Black may have helped him clarify his thinking. Certainly Black loaned Watt a considerable sum of money when the latter was struggling to develop his invention. On Black's death, Watt wrote '... to him I owe in great measure my being what I am; he taught me to reason and experiment in natural philosophy'.

Figure 4.3 Joseph Black (1728–1799)

In most cases, Hales threw away his airs after measuring their volume, so he cannot be credited with the discovery of gases such as oxygen, which must surely have been evolved in his experiments. Hales's great contribution was his apparatus. In the hands of later workers the tub of water became the pneumatic trough, and the suspended collecting vessel became a gas jar resting on a shelf in the trough.

GASES AS CHEMICAL ENTITIES

The first evidence of a gas being a distinct chemical entity and not merely a modification of atmospheric air was provided by Joseph Black (1728–1799). The origin of Black's work was an attempt to find a gentle but effective alkaline solvent for urinary calculi (stones), and in the course of his research Black unravelled the relationship between mild and caustic alkalis (carbonates and hydroxides). Black's paper, *Experiments on Magnesia Alba, Quick–lime and other alcaline substances*, was published in 1756.

Black's *Magnesia Alba* was basic magnesium carbonate, and he found that on strong heating seven–twelfths of its weight was lost. Only a small quantity of water could be condensed in a receiver, so the majority of the weight loss was caused by the escape of an air, which Black called *fixed air* (carbon dioxide). Black also found that the magnesia alba effervesced with dilute acids, but the product obtained on heating failed to do so. He then heated a weighed amount of magnesia, dissolved the solid residue in dilute sulphuric acid and treated the solution with mild alkali (potassium carbonate). He found that magnesia was precipitated, and that it was almost equal in weight to his starting material. He therefore demonstrated that fixed air was combined in mild alkali, and that it could be transferred to the residue obtained from heating magnesia, thus regenerating the original substance. In terms of modern equations:

$$MgCO_3 \rightarrow MgO + CO_2$$
$$MgO + H_2SO_4 \rightarrow MgSO_4 + H_2O$$
$$MgSO_4 + K_2CO_3 \rightarrow MgCO_3 + K_2SO_4$$

Black demonstrated that the production of quicklime by heating chalk was also accompanied by a weight loss, and that when quicklime was treated with a mild alkali the original chalk was formed along with a caustic alkali:

$$CaCO_3 \rightarrow CaO + CO_2$$
$$CaO + H_2O \rightarrow Ca(OH)_2$$
$$Ca(OH)_2 + K_2CO_3 \rightarrow CaCO_3 + 2KOH$$

Having demonstrated that magnesia and chalk lost fixed air on heating, and that this could be restored by the action of mild alkali, it remained to be shown that fixed air was a natural constituent of the atmosphere. It was well known that if lime water was exposed to the air it developed a crust of chalk, but Black established that lime water could be kept in corked bottles without developing any crust, and without the formation of any noticeable vacuum. Black therefore concluded that fixed air was not ordinary atmospheric air but a particular species of air which was dispersed throughout the atmosphere.

Black's work was of great importance for several reasons. By drawing attention to the role played by a particular gas in a series of chemical changes, he encouraged the development of pneumatic chemistry. His careful use of the balance and the deductions he made from the observed weight changes emphasised the importance of such measurements in chemical experimentation. Although he did not specifically state the law of conservation of mass, he tacitly assumed its validity. Perhaps most important of all, Black provided an explanation of a series of chemical changes without making any reference to the phlogiston theory, and he was one of those who provided Lavoisier with important clues when the latter was formulating his ideas (Chapter 5). Black's work also helped to establish the relationship between acids, alkalis and salts.

Black's work on alkalis had revealed some important properties of fixed air, and it was with this gas that the most famous of all pneumatic chemists, Joseph Priestley (1733–1804), commenced his researches around 1767. In that year he moved to the Mill–Hill chapel in Leeds, and by good fortune the chapel was close to a brewery, so he had an abundant supply of fixed air produced by fermentation. Priestley found that the gas would not support combustion and that mice soon died when placed in it, but that both the respirability of the gas and its ability to support combustion were improved by growing a plant in it. He also discovered that fixed air could be dissolved in water to produce what we would call carbonated or soda water. He recommended that brewers should manufacture soda water, and although he made no attempt to take any financial gain for himself, he became even more well known as a result of the invention.

Priestley soon extended his researches, and by modifying the technique of Hales for collecting gases he prepared and collected several distinct airs. Using familiar terminology, these were nitric oxide, nitrogen dioxide, nitrogen, nitrous oxide and carbon monoxide. Some of these gases were collected over mercury.

A contemporary of Priestley who made important contributions to pneumatic chemistry was Henry Cavendish (1731–1810). In 1766 he prepared inflammable air (hydrogen) by the action of dilute hydrochloric and sulphuric acids on zinc, iron and tin. Although Boyle had previously prepared inflammable air by a similar method (Chapter 3), Cavendish recognised that it was distinct from other inflammable gases by calling it the 'inflammable air from metals'. He thought this inflammable air was phlogiston, and since metals were supposed to consist of a calx united to phlogiston, it was reasonable to suppose that the acid was liberating phlogiston from the metal.

Priestley (Figure 4.4) was the son of a Calvinist weaver and cloth dresser, and he was born in the small stone cottage near Leeds where his father practised his craft. He received his higher education at the Nonconformist Daventry Academy. He then practised as a minister for six years during which time he started a school and purchased some scientific equipment, which included an air pump, an electrical machine and some microscopes. Such was his success as a teacher that in 1761 he was invited to join the Warrington Academy, another Nonconformist College, as tutor in languages and literature. Once at Warrington, Priestley suggested that Dr Matthew Turner of Liverpool be invited to give a course of lectures on chemistry. Priestley sat in on the lectures, and received his first instruction in chemistry alongside his own pupils. He decided to write a book on the history of electricity, and in the course of this task he repeated many experiments and discovered a number of new phenomena. Priestley achieved considerable fame as a result of the publication of his book, and in 1767 he accepted the position of minister of the Mill–Hill chapel in Leeds. This appointment enabled him to devote more time to scientific research, and he soon commenced his famous work on gases. In 1773 Priestley became 'literary companion' to the Earl of Shelburne, a politician who had been dismissed from his post for opposing the intransigent policy of George III towards the American colonies. Priestley's job was to help Shelburne formulate his opinions on current political issues, but he had ample time for his scientific work. When he left Shelburne's employment in 1780 Priestley moved to Birmingham, and with the aid of friends who backed him financially he was able to continue his scientific pursuits. His industry was astonishing and he wrote many articles, pamphlets and books on scientific, educational, religious, political and philosophical subjects. His scientific activities were closely linked to his religious convictions; he believed that through scientific enquiry man could understand the work of the Creator. He was opposed to the political and social systems of the day and supported the American and French Revolutions. He was regarded by some as a dangerous extremist, and his house and laboratory were ransacked during the riots in Birmingham on the second anniversary of the French Revolution in 1791. He then moved to London, but in 1794, when the French revolutionaries started sending their opponents to the guillotine and war broke out between Britain and France, Priestley again felt threatened and emigrated to the United States. He built a house at Northumberland, Pennsylvania, where he died in 1804. To the end he remained a staunch supporter of the phlogiston theory, which his own experiments had done so much to undermine.

THE DISCOVERY OF OXYGEN

When Priestley entered Shelburne's employment, he continued his work on gases (Figure 4.6) and isolated ammonia, sulphur dioxide and silicon tetrafluoride. His most famous discovery during this period was that of oxygen, which he first obtained in 1774. Priestley had been given a number of substances to investigate by his friend John Warltire, a travelling lecturer in natural philosophy. One of these materials was *mercurius calcinatus per se* (mercury(II) oxide). Priestley had recently acquired a lens of twelve inches in diameter, which enabled him to heat substances to a high temperature by focusing sunlight upon them. When *mercurius calcinatus per se* was subjected to this treatment, mercury was formed and oxygen was evolved.

Figure 4.4 Joseph Priestley (1733–1804)

Cavendish (Figure 4.5) was the son of Lord Charles Cavendish, who was the third son of the fifth Duke of Devonshire. As a young man, Henry received only a meagre allowance from his father. He acquired the habit of living very economically, and when he subsequently inherited several fortunes he continued to adopt a relatively modest existence. When he died he was the largest depositor at the Bank of England and was described as 'the richest of all the learned and the most learned of all the rich'. He was shy and awkward in company, and avoided entering into conversation, especially with women. He lived entirely for his scientific work. As well as his work in chemistry, he performed some fundamental researches in electricity and heat, most of which remained unpublished. Evidence of this work was found among his papers many years after his death. His death was as lonely as his life; sensing that the end was near, he instructed his servant to leave the room and not to come back until a prearranged time. When the servant returned, he found that his master was dead.

Figure 4.5 Henry Cavendish (1731–1810)

As Priestley subsequently remarked, he might well not have attempted the experiment had he considered the situation carefully beforehand. In terms of contemporary theory, mercury had already lost phlogiston when it had been heated to form *mercurius calcinatus per se*, and the evolution of anything else on heating to a higher temperature was totally unexpected. Even more remarkable was the behaviour of the new gas. As Priestley remarked: 'but what surprised me more than I can well express was, that a candle burnt in this air with a remarkably vigorous flame .... I was utterly at a loss how to account for it'. Priestley decided that the new gas was *dephlogisticated air* because it appeared to absorb phlogiston from a burning material more vigorously than naturally phlogisticated ordinary air. While he was able to account for the properties of oxygen reasonably satisfactorily in terms of the phlogiston theory, Priestley's explanation as to why the mercury calx should contain dephlogisticated air was far from convincing.

Priestley estimated that the new gas was 'five or six times as good as common air', and he breathed it himself and found that 'his breast felt peculiarly light and easy for some time afterwards'. In 1778 he found that aquatic plants growing in water containing dissolved fixed air evolve dephlogisticated air, and he commented: 'The injury which is continually done to the atmosphere by the respiration of such a large number of animals is, in part at least, repaired by the vegetable creation.' A few months after his discovery of oxygen, Priestley accompanied Shelburne to Paris where they met Lavoisier. Priestley described his findings to Lavoisier, who was at first sceptical, but he soon repeated Priestley's experiments and started to formulate his own views on combustion and respiration.

A pneumatic chemist whose claim to fame is at least as great as that of Black, Priestley or Cavendish is the Swede Carl Wilhelm Scheele (1742–1786). He concluded that the atmosphere was composed of a mixture of two elastic fluids (gases). In 1773 he allowed air to stand over liver of sulphur (potassium sulphide), which absorbed the oxygen. He found that the residual gas was slightly lighter than ordinary air, and since it did not support combustion he called it *foul air* (nitrogen). In keeping with current theory, he concluded that *foul air* had no ability to absorb phlogiston. The other component of air, which he estimated to be between one-third and one–quarter of the total, he deduced would have a particularly strong attraction for phlogiston, and he named it *fire air*. He then set about isolating fire air.

Scheele (Figure 4.7) came from a humble background and his family was unable to pay for him to receive a higher education. At the age of fourteen he was apprenticed to an apothecary, and he spent the rest of his life in apothecaries' shops, pursuing his researches in his spare time. He was a superb experimentalist, and prepared a large number of new substances. Among these were chlorine (which he isolated in 1774 and called dephlogisticated marine acid), and a considerable number of inorganic and organic acids. He was the first to observe the action of light on silver salts. His most important work was on the composition of atmospheric air and the preparation of oxygen, which he achieved before Priestley. Scheele prepared a number of highly toxic compounds, and his early death may have been due to his experiments with substances such as hydrogen cyanide.

Scheele subscribed to the current theory that heat was a material substance, and he proposed that it was composed of fire air combined with phlogiston. He therefore argued that he must expose to heat a substance that had a greater attraction for phlogiston than fire air. He decided to use nitric acid, and he heated nitre (potassium nitrate) with oil of vitriol (sulphuric acid), and after absorbing the brown fumes in lime water he found that the residual gas had the anticipated property of supporting combustion much more readily than air itself. Scheele subsequently prepared oxygen by several other methods, including heating mercury calx. Not only did Scheele's method of preparing oxygen pre–date Priestley's, but also his experiments were carefully planned and his fire air displayed the properties he had predicted for it. Unfortunately, Scheele's work was not published until 1777, by which time Priestley's work was well known.

Figure 4.6 Priestley's apparatus for experiments on gases

Figure 4.7 Carl Wilhelm Scheele (1742–1786)

THE COMPOSITION OF WATER

In 1781 Priestley performed some tests to see if heat (which was thought to be a material substance) had weight. These experiments led to a train of events that resulted in the composition of water being established. Until this time neither the synthesis nor the decomposition of water had been demonstrated, and ancient ideas of its elementary nature persisted. Priestley exploded inflammable air and common air in a weighed closed glass vessel by means of an electric spark, and he reweighed the apparatus after it had cooled down. In some of his experiments there appeared to be a loss of weight, and Priestley also noticed that the inside of the vessel 'became dewy'.

When Cavendish heard of these experiments, he repeated them, and found the loss of weight to be insignificant. He then turned his attention to the dew and found it to be pure water. In careful quantitative work he found that the common air was reduced in volume by up to one-fifth as a result of the explosion, and that this maximum diminution in volume occurred when the bulk of common air was two-and-a-half times that of inflammable air. He then performed further experiments in which he sparked inflammable air with Priestley's dephlogisticated air. He found that when the gases were present in a volume ratio of 2.02:1 they were completely converted to water.

Cavendish's earlier identification of inflammable air with phlogiston received support from an experiment performed by Priestley in 1782. He heated red lead in a vessel full of inflammable air by focusing the sun's rays upon it. Most of the inflammable air was absorbed, and metallic lead was formed. Thus the red lead (calx) appeared to be combining with the inflammable air, and since metals were supposed to consist of a calx combined with phlogiston, it looked as though the inflammable air must be phlogiston itself. However, Cavendish now believed that inflammable air was not simple phlogiston, but phlogiston combined with water. He also thought that dephlogisticated air was water deprived of its phlogiston. He therefore represented the reaction between the two as a transfer of phlogiston:

$$(\text{phlogiston} + \text{water}) \ + \ (\text{water} - \text{phlogiston}) \ \longrightarrow \ \text{water}$$
$$\textit{inflammable air} \qquad \textit{dephlogisticated air}$$

In 1783 Cavendish's assistant, Charles Blagden, visited Paris and told Lavoisier about Cavendish's work. Lavoisier rapidly repeated the experiments, and interpreted them in terms of his antiphlogistic theory by stating that water is a compound of inflammable air and oxygen (Chapter 5).

In some of the experiments in which Cavendish sparked inflammable air and dephlogisticated air, he found that the water produced contained a quantity of nitrous acid (this was the compound now known as nitric acid). Cavendish suspected that this had been formed by phlogisticated air (nitrogen) present as an impurity in the dephlogisticated air. Phlogisticated air had first been obtained in 1772 by a pupil of Black, Daniel Rutherford (1749–1819), who had obtained the gas by confining mice in air until it was no longer respirable, and then absorbing the fixed air with caustic potash. Cavendish sparked moist phlogisticated air and dephlogisticated air and found that nitric acid was indeed produced, thus showing that phlogisticated air (nitrogen) was a component of nitric acid. He decided to see how much phlogisticated air could be reacted, so after sparking for a while he added fresh oxygen and continued to spark, absorbing the nitric acid in caustic potash solution, until no further diminution in volume occurred. After absorbing the excess oxygen, he found that a small volume of unreacted gas remained. Cavendish estimated that the fraction of phlogisticated air that resisted reaction was about 1/120th. This remarkable observation remained unexplained for over a hundred years until Rayleigh and Ramsay showed that the unreactive gas contained a new inert element, which they named *argon* (Chapter 9).

For most of the eighteenth century the phlogiston theory was the principal theory of chemistry. As chemical knowledge increased, the theory was forced to adopt ever more convoluted explanations to account for the new discoveries. The pneumatic chemists provided the evidence upon which a new theory could be based. The chemist who was principally responsible for sweeping aside the old ideas and bringing about a revolution in chemical thought was Antoine Laurent Lavoisier.

5

Lavoisier and the Birth of Modern Chemistry

In the space of about 20 years commencing in 1770, the science of chemistry experienced a change more complete and more fundamental than any that had occurred before or has occurred since. Prior to 1770, almost all chemists accepted the doctrine of phlogiston as the great unifying principle of the science, and, if there were some facts that were hard to reconcile with the phlogiston theory, it was assumed that a modification or refinement of the theory would suffice. By 1790 not only had most chemists accepted the oxygen theory of combustion, but also the nomenclature of chemistry had been reformed, the modern concept of an element had been established, and there had appeared the first textbook to interpret chemistry in terms of the new ideas. All this was principally the work of one man, Antoine Laurent Lavoisier (1743–1794). It is one of the ironies of history that Lavoisier, who brought about what has been called *The Chemical Revolution*, was himself the innocent victim of a political revolution. He was sent to the guillotine while he was still at the height of his powers.

THE OVERTHROW OF AN ANCIENT THEORY

As a member of the Academy of Sciences, Lavoisier was asked to prepare reports on a wide range of matters of public interest. Either alone or in collaboration with others, he compiled over 200 such reports on topics that included the prisons of Paris (1780), aerostatic machines (balloons, 1784) and the hospitals of Paris (1787). In the late 1760s he was investigating the drinking water supplies of Paris, and this work was to lead him to conduct an important piece of fundamental research.

In the discussions Lavoisier had with other members of the Academy, it was clear that many still held the view that water could be transmuted into earth. The evidence for this assertion was van Helmont's tree experiment (Chapter 3), and the fact that even after several distillations water could be evaporated and still leave a solid deposit. Lavoisier set out to discover the true origin of the solid. He used a pelican, a piece of apparatus familiar to the alchemists, which enabled repeated distillation to occur within the same vessel (Figure 5.2). Having weighed the stoppered pelican empty, he added a quantity of rain water that had already been distilled eight times. He heated the pelican for a while, withdrawing the stopper occasionally to allow air to escape. He then inserted the stopper firmly, allowed the apparatus to cool, and reweighed. He then heated the tightly stoppered pelican on a sand bath so that continuous distillation occurred for a period of 101 days. After this time, solid material was clearly visible suspended in the water, so he cooled and weighed the apparatus once more. The weight was unchanged, and this result by itself disproved one theory concerning the origin of the solid material. This was that particles of fire, as proposed by Boyle (Chapter 3), had penetrated the vessel and combined with the water to produce earth.

Lavoisier (Figure 5.1) qualified for the legal profession but also studied mathematics, astronomy, botany, mineralogy, anatomy, meteorology and chemistry. As soon as his education was complete, he spent several years collaborating with one of his former teachers, Jean–Etienne Guettard (1715–1786), on the production of a geological map of France. His first original scientific work dates from this period and concerned the various forms of calcium sulphate. Lavoisier showed that gypsum loses water on heating, and that the resulting material (plaster of Paris) sets when mixed with water as a result of a recombination process. This paper was presented to the Academy of Sciences in 1765, and in the same year Lavoisier entered a competition organised by the Academy for an essay on the best means of lighting the streets of a large town at night. Lavoisier's essay was specially commended, and he was awarded a gold medal by the King. As a result of these activities, he was elected a member of the Academy in 1768 at the early age of twenty-five.

By this time Lavoisier had decided to devote his life to science. Although he had been left a considerable fortune by his mother, he wished to increase his wealth in order to pursue his scientific researches. Accordingly, he entered the *Ferme*, which was a company whose members purchased from the Government the privilege of collecting the national taxes. Clearly such an arrangement was open to abuse by the *Fermiers*, but there is no doubt that Lavoisier and his colleagues carried out their duties with complete fairness and honesty. Nevertheless, in the period following the French Revolution, the tax collectors of the previous regime were the subject of popular hatred, and it was Lavoisier's association with the Ferme which was to be the cause of his downfall.

In 1771, three years after he joined the Ferme, Lavoisier contracted an arranged marriage with Marie Anne Pierrette Paulze, the daughter of another Fermier. Lavoisier was 28 and his bride was only half his age, but later she was to become an invaluable partner in his scientific endeavours. After the wedding she studied languages, and she eventually acquired sufficient proficiency in English to translate scientific works for her husband. She assisted in his laboratory, and many of the entries in his laboratory notebook are in her hand. She studied with the painter Jacques–Louis David (1749–1825), who painted the well known portrait of her and her husband (Figure 5.1), and she drew the illustrations for her husband's textbook. She was a famous hostess, and scientists of all nationalities were welcomed by the Lavoisiers. Some of these visitors provided Lavoisier with vital information when he was formulating his new theory.

In 1775, Lavoisier was appointed to the Gunpowder Commission, and he soon improved both the quality and quantity of the powder produced. On his appointment he took up residence at the Arsenal, and there he set up a laboratory far superior to anything that had been established previously. After his arrest his property was confiscated, and the inventory showed that he possessed over 13 000 items of glassware and other chemical apparatus, and over 250 physical instruments, including three precision balances. From 1787 Lavoisier supervised an apprentice at the national gunpowder factory called Eleuthère Irénée Dupont de Nemours (1771–1834). Dupont was later to emigrate to the United States and start a gunpowder factory of his own. This became the nucleus of a major chemical company.

Lavoisier was a man of enormous energy and expended a great deal of effort on public causes. He made many improvements in the tax farm, he was a member of many boards and commissions, and he proposed many political, economic and social reforms. He established an experimental farm and made valuable suggestions as to how agricultural practices might be improved. Until the time of his arrest in 1793 he was a member of the Commission of Weights and Measures, which devised the system that developed into the modern metric system. But neither his scientific achievements nor his record of public service were sufficient to save him when the Fermiers were tried during the Terror that followed the French Revolution. The Revolutionary Tribunal was unable to find anything wrong with the way the Fermiers had discharged their duties, so they were found guilty on the absurd charge of plotting with the enemies of France. They were guillotined on 8 May 1794, and on that day Madame Lavoisier lost both her husband and her father.

Having established that nothing had entered the vessel, Lavoisier removed the contents and on weighing the dry pelican he found that there had been a decrease in its weight. He filtered and dried the solid suspended in the water, but its weight was only about one–quarter of the decrease in the weight of the pelican. Lavoisier suspected that the remaining material extracted from the glass was held in solution in the water, so he evaporated this to dryness and obtained more solid. The total weight of the solid material was now slightly greater than that lost by the pelican, which Lavoisier deduced was due to some solid being leached from the vessel used for the evaporation of the water. Lavoisier had thus demonstrated that water was not being converted into earth, and had accounted for the origin of the solid material formed on distillation.

EARLY WORK ON COMBUSTION

The year 1772 has been described as the crucial year for Lavoisier, for it was then that he started his experiments on combustion, which were to have such important consequences. His interest in the subject was aroused when he collaborated with some of his colleagues from the Academy on experiments which involved heating diamonds. They found that diamonds heated in a retort lost weight, but that if air was excluded the diamonds were unaffected. When diamonds were heated very strongly in the presence of air using a large burning lens (Figure 5.3), they were shown to be combustible.

Figure 5.1 Marie Anne Pierrette Lavoisier (1758–1836) and Antoine Laurent Lavoisier (1743–1794)

Figure 5.2 The pelican (left) was an apparatus for continuous distillation much used by the alchemists. It was named after the bird

Lavoisier, now working alone, turned his attention to the combustion of phosphorus and sulphur. He showed that air was necessary for the combustion of phosphorus and that the product of the reaction (called by Lavoisier phosphoric acid, actually phosphorus(V) oxide) weighed more than the original phosphorus. He also detected an increase in weight when sulphur was burnt.

Figure 5.3 The burning lens was made for the Academy of Sciences in 1774. A similar lens was used by Lavoisier in 1772 to heat diamonds and also a mixture of lead oxide and powdered charcoal

Lavoisier's compatriot, Louis Bertrand Guyton de Morveau (1737–1816), had recently confirmed that metals increase in weight on calcination. Although this phenomenon had been described by Boyle and others, doubt had frequently been expressed about both the reality and generality of the weight increase. It had been argued that the increase might be due to the addition of impurities from the vessel or the fuel. De Morveau's carefully conducted experiments excluded these possibilities, and he showed that a weight increase occurred on the calcination of copper, iron, tin, antimony, bismuth and zinc. Although de Morveau gave a conventional phlogistic explanation for his findings, Lavoisier speculated on whether air was involved in the calcination process. He knew that when a powdered calx was reduced to the metal it sometimes appeared to bubble or effervesce, and he suspected that this might be due to the evolution of air. He therefore determined to reduce a sample of lead oxide with charcoal using a burning lens to see if any air was evolved. Using a modification of Hales's apparatus (Chapter 4), he collected a large quantity of gas. On the basis of these experiments, Lavoisier formed his first tentative theory, that in all cases of combustion and calcination where an increase in weight is observed, fixation of air had occurred, and that when a calx is reduced, air is liberated. In order to establish his claim to be the originator of the new theory, he deposited a sealed note describing his ideas with the secretary of the Academy on 1 November 1772.

At this stage Lavoisier referred only to air and not to any particular part of it. He realised that a lot more detailed experimentation would be necessary and that his theory, if confirmed, would have far–reaching consequences. In a preface to his laboratory notebook, written in February 1773, Lavoisier stated:

'... The importance of the end in view prompted me to undertake all the work which seemed to me destined to bring about a revolution in physics and chemistry .... The results of the other authors whom I have named, considered from this point of view, appeared to me like separate pieces of a great chain; these authors have joined only some links of the chain. But an immense series of experiments remains to be made in order to lead to a continuous whole'.

Among the other authors referred to by Lavoisier were Hales, Black and Priestley. There followed a period of rapid experimentation, leading to the publication, in January 1774, of Lavoisier's first book, *Opuscules Physiques et Chymiques*, in which he described his results to date. He had established that between one–fifth and one–sixth of the atmospheric air was absorbed in the combustion of phosphorus, and that the gas evolved when lead oxide was reduced with charcoal was identical to Black's fixed air. He was, however, uncertain as to whether it was fixed air that was absorbed in combustion and calcination, or whether it was common air or some other gas contained in it.

THE OXYGEN THEORY

Lavoisier then repeated Boyle's experiments on the calcination of tin and lead (Chapter 3), but with an important addition to the procedure. A known weight of tin was placed in a weighed retort, which was then sealed. The retort was heated for two hours and then allowed to cool. At this stage, Lavoisier weighed the retort, which Boyle had omitted to do. Lavoisier found the sealed retort had not increased in weight during the experiment, and thereby disproved Boyle's theory that the increase in weight was due to the passage of fire particles through the glass. When he opened the retort, Lavoisier heard air rush in. The opened retort was now heavier than previously, and the increase in weight caused by the admission of air was almost exactly equal to the observed increase in weight of the tin. The fact that the calcination of the tin appeared to cease after one hour, and that the increase in weight on opening the retort was considerably less than it would have been if a vacuum had been created inside, confirmed that only a portion of the air was absorbed in the calcination.

It was at this point, in October 1774, that Priestley visited Lavoisier and described the remarkable air he had obtained from mercury calx some two months previously. As we have seen (Chapter 4), Priestley was surprised with the result of his experiment, but Lavoisier was already convinced that a gaseous material was removed when a calx was reduced to the metal. Even before Priestley performed his experiment, Lavoisier's compatriot, Pierre Bayen (1725–1798), had shown that charcoal was unnecessary for the reduction of mercury calx; strong heat alone was sufficient. Furthermore, he found that there was a decrease in weight during the reduction. Lavoisier repeated Priestley's experiment, and he also reduced the mercury calx by heating it with charcoal. He demonstrated that fixed air was produced in the latter experiment, but that the *eminently respirable air* (Lavoisier's term for Priestley's dephlogisticated air, i.e. oxygen), which was produced when mercury calx was heated alone, had very different properties.

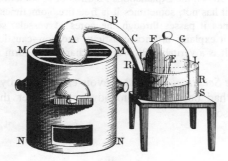

Figure 5.4 The apparatus used by Lavoisier to form the calx of mercury by heating the metal in air

Lavoisier's most famous experiment involved both the formation and decomposition of the mercury calx. A weighed quantity of mercury was placed in a large retort, which was connected to a bell jar of air standing over mercury in a trough (Figure 5.4). The mercury in the retort was kept at a temperature near its boiling point for twelve days, by which time the quantity of mercury calx on the surface of the metal had ceased to increase. On cooling, it was found that the air in the apparatus was reduced to five–sixths of its former volume, and that it was now fit for neither combustion nor respiration. Lavoisier termed this air *mephitic air*. When the mercury calx produced was heated more strongly in a small retort, a quantity of eminently respirable air was liberated equal in volume to that lost in the formation of the calx. When mephitic air and eminently respirable air were mixed in the appropriate proportions, the resulting mixture was identical to atmospheric air.

Lavoisier had therefore demonstrated that air was a mixture of two components, and that one of them, eminently respirable air, combined with metals during calcination and was the cause of the increase in weight. In further experiments, he showed that it was also eminently respirable air that was consumed in the combustion of carbon, phosphorus and sulphur. He showed that Black's fixed air was composed of carbon and eminently respirable air, and he thus accounted for the formation of fixed air when a calx was reduced using charcoal.

Lavoisier found that the products of the combustions of carbon, phosphorus and sulphur when dissolved in water were acidic, and he also demonstrated that eminently respirable air was combined in nitric acid. He concluded that eminently respirable air was an essential component of acids, and in 1779 he renamed it *principe oxygine*, the latter word being derived from Greek and meaning *acid begetter*. All acids therefore consisted of two parts: *principe oxygine* and a *radical*.

Lavoisier originally presented his new theory as an alternative to the phlogiston theory, but by 1783 he felt sufficiently confident to attack the old theory in a paper read to the Academy called *Reflections on Phlogiston*. In it he said:

> '... chemists have made phlogiston a vague principle which is not strictly defined and which consequently fits all the explanations required of it; sometimes the principle has weight, sometimes it has not; sometimes it is free fire, sometimes it is fire combined with earth; sometimes it passes through the pores of vessels, sometimes these are impenetrable to it; it explains at once causticity and non–causticity, colour and the absence of colour. It is a veritable Proteus that changes its form at every instant'.

Like Boyle in *The Sceptical Chymist*, Lavoisier had destroyed an earlier theory, but whereas Boyle had been unable to advance a better theory, Lavoisier had succeeded in doing so.

THE PROBLEM OF INFLAMMABLE AIR

One serious difficulty still remained for Lavoisier's theory. This concerned the inflammable air which Cavendish had described in 1766. To many adherents of the old theory, inflammable air was phlogiston itself. When an acid reacted with a metal to produce inflammable air, the phlogistonists represented the process thus:

$$\text{(calx + phlogiston)} + \text{acid} \rightarrow \text{(calx + acid)} + \text{phlogiston}$$
$$\qquad\quad \textit{metal} \qquad\qquad\qquad\qquad \textit{salt} \qquad\quad \textit{inflammable air}$$

Lavoisier was faced with the problem that if a metal was not composed of a calx combined with phlogiston, what was the origin of the inflammable air when a metal was reacted with an acid? Furthermore, the phlogiston theory gave a perfect explanation of Priestley's observation that a calx was converted to the metal on heating in inflammable air, which was absorbed in the process (Chapter 4):

$$\text{calx} + \text{phlogiston} \rightarrow \text{(calx} + \text{phlogiston)}$$
$$\qquad \textit{inflammable air} \qquad\qquad \textit{metal}$$

Lavoisier had also performed experiments in which he burned inflammable air in oxygen. He failed to detect any sign of acidity in the product, which was predicted on his theory of oxygen being the acid begetter.

Once again Lavoisier received a hint from a traveller from England. This time it was Charles Blagden, Cavendish's assistant, who visited Lavoisier in June 1783 and described Cavendish's quantitative experiments in which he had reacted two volumes of inflammable air with one of dephlogisticated air and found that in this ratio the mixture was completely converted into water. Cavendish gave a phlogistic interpretation of this result, but Lavoisier realised that the reaction represented a synthesis of water from simpler materials. Lavoisier repeated Cavendish's work in a rather hurried fashion, and also performed the important complementary experiment in which he decomposed water by passing it drop by drop along a heated copper tube containing iron to obtain iron calx and inflammable air. Lavoisier was then able to assert that water was not an element but a compound of oxygen and inflammable air, which was soon renamed hydrogen (water begetter).

Lavoisier now explained the production of hydrogen from the reaction of a metal with an acid by saying that the metal combined with the oxygen from the water liberating hydrogen, and the metal oxide immediately combined with the acid to produce a salt:

$$\text{metal} + \underset{water}{(\text{hydrogen} + \text{oxygen})} \rightarrow \underset{calx}{\text{metal oxide}} + \text{hydrogen}$$

$$\underset{calx}{\text{metal oxide}} + \text{acid} \rightarrow \text{salt}$$

Lavoisier regarded non-metal oxides as acids; today, it is the products of their reactions with water that are termed *acids*, and in the reaction with metals the hydrogen comes from the acid.

THE REFORM OF CHEMICAL NOMENCLATURE

Once these difficulties had been overcome, the new theory began to gain ground. Joseph Black was an early convert, and he was teaching the antiphlogistic chemistry to his students at Edinburgh before 1784. The French chemists de Morveau, Claude Louis Berthollet (1748–1822) and Antoine François de Fourcroy (1755–1809) accepted the theory in 1785 or soon after, and once won over they collaborated with Lavoisier in the reform of chemical nomenclature.

Chemistry was encumbered with a plethora of names surviving from the days of alchemy, such as *powder of Algaroth, liver of sulphur* and *butter of arsenic*. Furthermore, some of the recently discovered gases had been given names in accordance with the phlogiston theory, for example *phlogisticated air* (nitrogen). De Morveau had suggested in 1782 that chemical names should be reformed in a manner similar to that employed by Linnaeus in his systematisation of botanical nomenclature. De Morveau had been influenced by the Swedish chemist Torbern Bergman, who had been a student of Linnaeus. De Morveau proposed that a substance should have one fixed name, which should reflect its composition if known, and that names should generally be chosen from Greek or Latin roots. The new proposals of Lavoisier, de Morveau and their colleagues contained a list of *substances not decomposed*, and the composition of each compound was expressed in terms of these substances. The term *element* was not employed at this stage, but Lavoisier used it in his textbook published two years later. The new proposals were first presented to the Academy, and were then published by Lavoisier and his associates in 1787 under the title *Méthode de Nomenclature Chimique*.

Under the new system, calxes became oxides, oil of vitriol became sulphuric acid and the other oxyacid of sulphur was called sulphurous acid. The correspon- ding salts became the sulphates and the sulphites. The new nomenclature helped to publicise the new antiphlogistic chemistry; everyone who wanted to read a paper that used the new nomenclature had to think in terms of Lavoisier's oxygen theory. In spite of recent revisions in chemical nomenclature, most names proposed by Lavoisier and his associates are comprehensible to the chemist of today.

It is not surprising that it took longer for the new chemistry to be accepted by many outside France. The best defence of the old theory was *An Essay on Phlogiston* by the Irish chemist Richard Kirwan (1733–1812). This was published in 1787, and a French translation by Madame Lavoisier appeared the following year. The French translation contained a note at the end of each chapter by one of the French chemists refuting Kirwan's arguments. These notes were written by Lavoisier, de Morveau, Laplace, Monge, Berthollet and Fourcroy. The *Essay*, with a translation of the French notes, appeared in a second English edition in 1789. Although Kirwan made an attempt to defend his position in some notes of his own which he appended to the second English edition, in 1791 he too accepted the antiphlogistic chemistry.

LAVOISIER'S TEXTBOOK

Soon after the publication of the *Nomenclature*, Lavoisier decided to write another book to explain the new system more fully. However, the scope of the work gradually widened, as Lavoisier explained in the Preface:

> ' Thus, while I thought myself employed in forming a Nomenclature, and while I proposed to myself nothing more than to improve the chemical language, my work transformed itself by degrees, without my being able to prevent it, into a treatise upon the Elements of Chemistry'.

The book was published in 1789 as *Traité élementaire de Chimie* and was soon translated into other European languages, appearing in English under the title *Elements of Chemistry*. The appearance of the *Traité* is a landmark in the history of chemistry; its importance is comparable with that of Newton's *Principia* in the realm of mechanics. It was aimed at those studying chemistry for the first time, but the new chemistry was so different from its predecessor that most of Lavoisier's readers were beginners. The book did much to popularise the new antiphlogistic chemistry, and in 1791 Lavoisier wrote: 'All young chemists adopt the theory and from that I conclude that the revolution in chemistry is come to pass'.

The *Traité* completely broke with the tradition of earlier chemistry textbooks, which had mainly provided a series of descriptions of the preparation of a great many substances. The *Traité* was divided into three parts. The first was concerned with aeriform fluids (gases), and with combustion and the formation of acids. The second part dealt with the compounds formed by pairs of elements and the salts formed by the inorganic and organic acids known at the time. The third part was a description of the instruments and operations of chemistry. Lavoisier decided to omit any consideration of affinities or elective attractions on the grounds that 'the principal data are still wanting'.

In the *Traité* Lavoisier emphasised the importance of experimental evidence, and stressed the need to rid chemistry of ideas that could not be supported by such evidence: '... I have imposed upon myself the law of never advancing but from the known to the unknown, of deducing no consequence that is not immediately derived from experiment and observations'. Accordingly he was dismissive of the four–element theory of the Greeks: 'The notion of four elements, which, by the variety of their proportions, compose all the known substances in nature, is a mere hypothesis, assumed long before the first principles of experimental philosophy or of chemistry had any existence'.

Lavoisier introduced the modern definition of an element, namely that all substances that had not yet been decomposed should be accepted as elements. The table of elements given by Lavoisier (Figure 5.5) included earths such as lime and magnesia, but Lavoisier suspected that they were compounds and commented that experiment would probably soon show them to be so. He omitted the alkalis potash and soda from the list because 'these substances are evidently compound, although we do not know as yet the nature of the principles that enter into their composition'.

The list of elements was considerably shorter than the list of *substances not decomposed* given in the *Nomenclature*, which had listed the radicals of nineteen organic acids. Lavoisier now recognised that these were all composed of carbon and hydrogen. The idea of a compound radical as a group of atoms behaving as a unit became an important concept in nineteenth century organic chemistry.

Not surprisingly, some of Lavoisier's ideas were destined to be challenged and ultimately rejected. Central to his views was that oxygen was an essential constituent of all acids. He included in his table of simple substances the unknown element *muriatic radical*, which he deduced to be combined with oxygen in muriatic (hydrochloric) acid. Chlorine itself, discovered by Scheele in 1774, he called oxygenated muriatic acid. In 1809, Gay–Lussac and Thenard showed that muriatic acid gas contained no oxygen, and the following year Davy showed that Lavoisier's oxygenated muriatic acid was an element (Chapter 7).

Lavoisier included in his table of simple substances light and caloric (heat). In the eighteenth century the modern concept of energy lay far in the future, and heat and light were accepted as material substances. Lavoisier suggested that the difference between solids, liquids and gases depended upon the quantity of caloric present. He proposed that, on heating a solid, caloric particles surrounded the solid particles, thereby producing a liquid, and that if sufficient caloric was present a gas was formed. Thus oxygen gas was a compound of elemental oxygen and caloric. Here Lavoisier owed something to Black's discovery of latent heat.

In the *Traité* Lavoisier described an ice calorimeter, which he had devised in conjunction with Pierre Simon de Laplace (1749–1827) to measure the amount of caloric evolved in chemical changes. These experiments laid the foundations of thermochemistry (Chapter 13).

	Noms nouveaux.	Noms anciens correspondans.
	Lumière	Lumière.
Substances simples qui appartiennent aux trois règnes, & qu'on peut regarder comme les élémens des corps.	Calorique	Chaleur. Principe de la chaleur. Fluide igné. Feu. Matière du feu & de la chaleur.
	Oxygène	Air déphlogistiqué. Air empiréal. Air vital. Base de l'air vital.
	Azote	Gaz phlogistiqué. Mofète. Base de la mofète.
	Hydrogène	Gaz inflammable. Base du gaz inflammable.
Substances simples non métalliques oxidables & acidifiables.	Soufre	Soufre.
	Phosphore	Phosphore.
	Carbone	Charbon pur.
	Radical muriatique . .	Inconnu.
	Radical fluorique . . .	Inconnu.
	Radical boracique . .	Inconnu.
Substances simples métalliques oxidables & acidifiables.	Antimoine	Antimoine.
	Argent	Argent.
	Arsenic	Arsenic.
	Bismuth	Bismuth.
	Cobalt	Cobalt.
	Cuivre	Cuivre.
	Etain	Etain.
	Fer	Fer.
	Manganèse	Manganèse.
	Mercure	Mercure.
	Molybdène	Molybdène.
	Nickel	Nickel.
	Or	Or.
	Platine	Platine.
	Plomb	Plomb.
	Tungstène	Tungstène.
	Zinc	Zinc.
Substances simples salifiables terreuses.	Chaux	Terre calcaire , chaux.
	Magnésie	Magnésie , base du sel d'epsom.
	Baryte	Barote , terre pesante.
	Alumine	Argile, terre de l'alun, base de l'alun.
	Silice	Terre siliceuse, terre vitrifiable.

Figure 5.5 The table of elements from Lavoisier's *Traité élementaire de Chimie*

Another important feature of the *Traité* was that it contained the first explicit statement of the law of conservation of mass. In the chapter on fermentation Lavoisier wrote: 'We may lay it down as an incontestable axiom, that, in all operations of art and nature, nothing is created; an equal quantity of matter exists both before and after the experiment; the quality and quantity of the elements remain precisely the same; and nothing takes place beyond changes and modifications in the combination of these elements'. Lavoisier described a fermentation experiment in which he started with weighed quantities of sugar, water and yeast. He found at the end of the fermentation that the quantity of water was unchanged, and he weighed the residual sugar and yeast, and the carbonic acid (carbon dioxide), alcohol and acetic acid formed. Not only did he show that no matter had been created or destroyed, but also from his estimate of the elemental composition of each substance, he was able to show that the mass of each element remained unchanged. He was the first to estimate the percentage of each element in an organic compound, and his method, though relatively crude, formed the basis of later and more sophisticated techniques (Chapter 8).

The year 1789 not only saw the publication of the *Traité* but also marked the start of the French Revolution. Although Lavoisier was to lose his life in the Terror, there was one immediate and happier consequence. This arose from the freedom of the press that the Revolution brought about, and enabled eight chemists, including Lavoisier, to edit a new journal, the *Annales de Chimie*. This provided a medium for reports on the new chemistry, and although publication ceased after Lavoisier's arrest in 1793, it was resumed in 1797 and has continued until the present day.

Like other great men, Lavoisier has had his share of detractors. It has been pointed out that he discovered no new substances, and that some of the crucial pieces of evidence for his antiphlogistic theory were provided by others. However, it was he alone who abandoned the preconceptions of his predecessors and contemporaries and interpreted the experimental data afresh.

FRENCH CHEMISTRY AFTER LAVOISIER

Although the French Revolution resulted in the death of the nation's leading chemist, the overall effect of the Revolution on French science was probably beneficial. Lavoisier had demonstrated the service that a scientist could render the state, and the new authorities encouraged the education of scientists by establishing the École Polytechnique in 1794. Many of the leading French scientists of the first half of the nineteenth century were graduates of this institution. Napoleon was aware of the importance of science to the nation, and after he became Emperor in 1804 he did much to encourage science and scientists.

After the death of Lavoisier, the two leading figures in the French scientific community were Berthollet and Laplace. Both had participated in the informal gatherings that had taken place at Lavoisier's home at the Arsenal at which the new antiphlogistic chemistry had been discussed. In the introduction to the *Traité* Lavoisier acknowledged the importance of these discussions, and mentioned Berthollet and Laplace as participants.

Apart from the thermochemical work that he had performed with Lavoisier, Laplace's chief contributions to science were in the areas of mathematics, physics and astronomy. His five-volume *Mécanique Celeste*, published between 1799 and 1825, was the most important contribution in this field since Newton's *Principia*. Berthollet was a chemist and his most important work prior to the death of

Lavoisier was on dyeing and bleaching. In 1786 he had discovered that oxymuriatic acid (chlorine) could be used to bleach cloth. Scheele, who had discovered chlorine twelve years previously, seems only to have discovered its ability to decolorise paper, plants and flowers. Berthollet developed a satisfactory commercial bleaching process, but made no attempt to profit financially from the method he had devised.

When Napoleon embarked on his expedition to Egypt in 1798, he took a number of scientists with him, among them Berthollet. While in Egypt, Berthollet made the important observation that a chemical reaction might proceed in the reverse direction under appropriate conditions. He observed deposits of soda on the shores of Egyptian salt lakes, and concluded that the soda was produced as a result of a reaction between salt and limestone:

$$2NaCl + CaCO_3 \rightarrow Na_2CO_3 + CaCl_2$$

This reaction normally proceeds in the opposite direction, and Berthollet suggested that the extremely high concentration of sodium chloride was responsible for the reversal. Berthollet was here suggesting that the normal affinities of substances could be altered by the masses present; he was anticipating the law of mass action of Guldberg and Waage (Chapter 13). However, Berthollet also speculated that a compound could be of a variable composition determined by the masses of reactants employed in its preparation. Berthollet and Proust became engaged in a controversy on this issue (Chapter 6), and after 1808 most chemists sided with Proust and rejected Berthollet's views. An unfortunate consequence of this was that for many years most chemists dismissed any notion of mass action.

Figure 5.6 Joseph Louis Gay–Lussac (1778–1850)

Berthollet returned to France with Napoleon in 1799, and purchased a country house just outside Paris at Arcueil. There, Berthollet established a chemical laboratory, where he allowed promising young chemists to work. Berthollet's most famous protégé was Joseph Louis Gay–Lussac (1778–1850), who commenced work at Arcueil in 1801. In 1806 Laplace also bought a house at Arcueil, and he and Berthollet acted as the nucleus of a group of scientists who met at Berthollet's house and who were known as the *Society of Arcueil*. The Society flourished between 1807 and 1813, and used to meet at irregular intervals on Sundays, with Berthollet and Laplace alternating as chairman. There were never more than twelve members of the Society at any one time, and, apart from Berthollet, Laplace and Gay–Lussac, some of those who made important contributions to chemistry were Louis Jacques Thenard (1777–1857), Pierre Louis Dulong (1785–1838) and Jean–Baptiste Biot (1774–1862). The Society published a journal comprising the papers read at the meetings, and three volumes appeared before the Society ceased to exist.

Gay–Lussac (Figure 5.6) was the most famous of the protégés of Berthollet. After distinguishing himself at the École Polytechnique, Gay–Lussac went to Arcueil in 1801 to work as Berthollet's assistant, and Berthollet was to play an important role in Gay–Lussac's professional advancement. Gay–Lussac's first major research was on the thermal expansion of gases (Chapter 13), and his work was considerably more accurate than a similar study performed at about the same time by Dalton.

In 1804 Gay–Lussac made two balloon ascents and conducted a number of scientific experiments at high altitude. On his second ascent, Gay–Lussac reached a height of over 7000 m, a record that stood for 50 years. He took samples of air at an altitude of over 6000 m, and his subsequent analyses revealed that the proportion of oxygen was the same as at sea level. The method of analysis employed was to use the eudiometer devised by Volta, in which the air was sparked with hydrogen to produce water vapour which condensed. The contraction in volume enabled the proportion of oxygen in the air to be calculated. This method clearly depended upon the combining ratio of hydrogen and oxygen being known accurately, and soon after the balloon ascent Gay–Lussac collaborated with the German explorer Alexander von Humboldt (1769–1859) in the redetermination of this ratio. Two years previously Humboldt had climbed to within 500 m of the summit of Chimborazo in the Andes, then thought to be the highest mountain in the world, and like Gay–Lussac he had taken samples of air at high altitude. Gay–Lussac and Humboldt found that 100 parts by volume of oxygen combined with 199.89 parts of hydrogen. Gay–Lussac was to return to the study of the combining volumes of gases three years later, as a result of which he announced his famous law (Chapter 6).

The sum total of Gay–Lussac's chemical achievements is enormous. Either alone or in collaboration with Louis Jacques Thenard (1777–1857) he worked on many topics. These included the alkali metals, the nature of chlorine, the properties of iodine, and the nature of acids (he showed conclusively that some acids contain no oxygen and introduced the term hydrochloric acid for muriatic acid). He considerably improved upon Lavoisier's method for the analysis of organic compounds by introducing an oxidising agent. It was Gay–Lussac who was responsible for developing volumetric analysis into a widely useful and applicable technique. He held academic appointments in both chemistry and physics, and not the least of his claims to fame was that he allowed the young Liebig to work in his laboratory as his assistant.

It is probable that, at Arcueil, Berthollet and Laplace were trying to recreate a stimulating atmosphere such as had existed at Lavoisier's home at the Arsenal. Whether this is true or not, the magnitude of Lavoisier's chemical legacy is not in doubt, and his influence is evident in much of the work done in the early decades of the nineteenth century. Phlogiston was overthrown, and with a new working definition of an element it was not to be long before new ideas were being formulated about the fundamental particles of matter.

6

The Chemical Atom

The idea that all matter is composed of minute atoms had first been suggested in ancient Greece, and Boyle had employed the concept of atomism in his mechanical philosophy. But atomism had as yet had little success in explaining the phenomena of chemistry. It was difficult to see how these minute particles, which were supposed to be the universal building blocks of nature, could provide an explanation for the chemical properties of the huge number of different materials known, a number that was continually increasing.

Lavoisier's concept of an element provided the foundation for the new version of atomism which appeared in the early years of the nineteenth century. The new theory, known as the *chemical atomic theory*, had as its central tenet that different elements have fundamentally different atoms. In consequence, repeated division of a piece of lead would ultimately yield the last atom of lead, and similarly repeated division of a piece of gold would yield the last atom of gold. If lead and gold were genuine elements, then the atoms of lead and the atoms of gold were different. This explained the different properties of lead and gold, and the transmutation so keenly sought by the alchemists became a theoretical impossibility.

Working in the tradition of Black and Cavendish, Lavoisier had emphasised the importance of the use of the balance in chemical investigations. The balance had been employed by assayers and mineral prospectors for hundreds of years, but in Lavoisier's time the precision balance was still a specialised piece of equipment. Black had used an apothecary's balance, but Cavendish and Lavoisier had much more accurate balances constructed to their own specifications. The age of quantitative chemistry had now truly arrived.

QUANTITATIVE ANALYSIS AND THE LAW OF CONSTANT COMPOSITION

The techniques of quantitative analysis were advanced significantly by the German, Martin Klaproth (1743–1817). He devoted much attention to sources of error and their elimination. He published all his experimental data, such as weights of samples and precipitates, so that others could check for any errors. Frequently in the analysis of minerals he found that the sum of the constituents was less than 100 per cent. Previous workers had tended to ignore such discrepancies, ascribing them to experimental error, but Klaproth realised that in such cases there might be a previously unknown substance present. In 1789 he thus discovered the oxides of the hitherto unknown elements uranium and zirconium, although it was many years before the metals themselves were isolated. Another famous analyst of the period was Louis Nicolas Vauquelin (1763–1829), who discovered chromium in 1798. Around this time, several other metals were discovered by the careful analysis of minerals.

Quantitative analysis was soon employed to verify the law of constant composition. It had for many years been assumed by most chemists that different samples of the same pure compound would have identical compositions. In 1799, the Frenchman Joseph Proust (1754–1826) analysed basic copper carbonate of natural occurrence (malachite) and the same compound prepared in the laboratory. Both samples gave the same analytical results. He demonstrated a similar constancy of composition in many other compounds, and he showed that several metals form more than one oxide and sulphide, each of definite composition.

Proust's conclusions were immediately challenged by Berthollet, who maintained that many compounds could have a variable composition. Berthollet quoted the example of the metal copper, which appeared to form a wide range of oxides. Proust pointed out that this was due to the formation of different mixtures of two oxides, each of definite composition. Berthollet's concept of continuously variable composition would have been hard to reconcile with the chemical atomic theory, but by about 1808 Proust's views were generally accepted. However, many years later it was discovered that some compounds, such as the oxides and sulphides of iron, could indeed have a variable composition. In the case of iron(II) sulphide, a compound corresponding to the formula FeS is rarely encountered, and samples are usually deficient in iron to a variable extent. This is due to some of the lattice sites of iron(II) ions being vacant, while others are occupied by iron(III) ions to maintain electrical neutrality. These non–stoichiometric compounds are sometimes called *Berthollide compounds*.

Quantitative results of a different kind were obtained by Jeremias Richter (1762–1807). He was obsessed with obtaining mathematical relationships in chemistry, and he helped to establish the concept of equivalent or combining weight. His work was summarised in 1802 by Ernst Fischer (1754–1831), who produced a table of equivalent weights of acids and bases related to sulphuric acid having a value of 1000. On this scale, muriatic acid (HCl) had a value of 712, and soda and potash had values of 859 and 1605 respectively. This meant that 859 parts of soda or 1605 parts of potash were required to neutralise 1000 parts of sulphuric acid or 712 parts of muriatic acid.

To the modern mind, ideas such as definite proportions and equivalent weight seem to cry out for an explanation in terms of a chemical atomic theory. However, we must remember that today everyone is taught to think in terms of atoms and molecules very early in their study of chemistry. Such explanations were certainly not obvious at the beginning of the nineteenth century, and indeed John Dalton

(1766–1844), who propounded the chemical atomic theory, started to think about atoms as a result of a study of phenomena which were physical rather than chemical.

Dalton (Figure 6.1) was born in Eaglesfield, near the English Lake District. His father was a hand loom weaver, and, since the family were Quakers, Dalton was educated at the local Quaker schools. At the age of twelve he opened his own Quaker school, but he had little authority over some of his rougher pupils, and he soon relinquished his post to work as a farm-hand. At the age of fifteen he moved to Kendal to assist at a Quaker boarding school where his brother Jonathan was already teaching, and four years later the brothers were in charge of the school.

At Kendal, Dalton was befriended by John Gough. Dalton learned much from this remarkable man, who although blind was an expert in Latin, Greek, mathematics and science. It was Gough who encouraged Dalton to start keeping a meteorological journal. His records commenced on 24 March 1787, and were made every day until the day before he died 57 years later.

For a time Dalton considered entering the medical profession. However, his reputation as a teacher was growing, and he was giving lectures on scientific subjects to the general public in Kendal. Consequently he was invited in 1793 to teach at the New College in Manchester, a Dissenting Academy similar to the one in nearby Warrington which had employed Priestley. Soon after his arrival in Manchester, Dalton published his first book, entitled *Meteorological Observations and Essays*. This was written in Kendal, and reveals Dalton's lack of a formal scientific education and his lack of contact with fellow scientists. Nevertheless, the book contains Dalton's first speculations concerning the atmosphere and the water vapour it contains, and it was as a result of such speculations that the chemical atomic theory was born.

Dalton's move to Manchester ended his isolation, for he immediately joined the Literary and Philosophical Society, which had been founded in 1781. The first paper he read before the Society was concerned with the colour blindness from which he suffered, and was the first description of this syndrome, which is still occasionally called *Daltonism*. Membership of the Society introduced Dalton to a circle of friends with whom he was able to discuss the ideas that had been forming in his mind concerning the atmosphere. Particularly important was his friendship with William Henry (1774–1836), who had worked on the solubility of gases in water.

After the publication of his atomic theory, Dalton remained in Manchester, defending the theory and seeking further experimental evidence for it. In spite of the international reputation he eventually acquired, he continued to work as a teacher. In 1800 he had resigned from New College and he concentrated on giving private tuition in mathematics, experimental philosophy and chemistry. He also gave courses of public lectures in Manchester and other cities. When he died he was widely regarded as Manchester's leading citizen.

DALTON'S THEORIES ON GASEOUS MIXTURES

Initially, Dalton employed a model for a gas very similar to that used by Newton. It is important to realise that Newton's (and Dalton's) picture of a gas was essentially a static one, and quite different from the dynamic model provided by the kinetic theory of gases, which was not developed until the middle of the nineteenth century. Newton imagined a gas to be a three–dimensional array of mutually repulsive particles. He showed that, if the repulsive force is inversely proportional to the distance between the particles, then the pressure of a fixed quantity of gas doubles when the volume is halved, in accordance with Boyle's law.

Figure 6.1 John Dalton (1766–1844)

Newton had imagined all air particles to be identical, but by Dalton's time it had been shown that air contains both oxygen and nitrogen, with variable amounts of water vapour. In 1801 Dalton rejected the current idea that air was a loose chemical compound between these constituents, but was then confronted with the problem that a physical mixture should separate into layers, with the most dense substance (oxygen) forming the bottom layer. In an attempt to explain the homogeneity of air, Dalton hit upon the idea that each particle could only repel others of its own kind, and that dissimilar particles exerted no forces on each other. From this he deduced that, in a mixture of gases, each constituent should exert the same pressure as it would if it alone occupied that volume. This conclusion, although based on an incorrect model, is valid, and is now known as Dalton's law of partial pressures.

Dalton soon became dissatisfied with this model for gaseous mixtures, for it was difficult to imagine how particles could repel only those of their own kind. Dalton considered the gaseous particles to consist of hard centres surrounded by an atmosphere of heat (caloric). The atmosphere of heat was supposed to be the cause of the repulsive power, so on this basis there should be repulsion between dissimilar particles. Dalton's next idea was that different types of particle had different sizes, and that in consequence no equilibrium could be established between particles of different sizes exerting different forces. Therefore, the constituents of a gaseous mixture would not separate into layers. If the particles of different gases had different sizes, they would also presumably have different weights.

Dalton first announced his theory concerning the sizes and weights of particles of different gases in a paper concerned with the solubility of gases in water. This paper was read to the Manchester Literary and Philosophical Society in 1803 and published in 1805. Dalton had performed experiments of his own on the subject, and he also used the results obtained by his friend William Henry, who had discovered that the mass of gas absorbed by a liquid is proportional to the pressure. Dalton argued that, when a gas dissolves in a liquid, the process is simply one of mechanical mixing of the gaseous particles with those of the liquid, and is not a chemical combination. He then admitted that the greatest weakness of the mechanical hypothesis is that a liquid like water should dissolve all gases to the same extent, but stated that this difficulty is overcome if the ultimate particles of different gases have different sizes and weights. He continued:

'An enquiry into the relative weights of the ultimate particles of bodies is a subject, as far as I know, entirely new: I have lately been prosecuting this enquiry with remarkable success. The principle cannot be entered upon in this paper; but I shall just subjoin the results, as far as they appear to be ascertained by my experiments'.

There then follows the first table of atomic weights ever published (Table 6.1). Dalton does not reveal here how his atomic weights were calculated, and tantalisingly he does not elaborate his theory any further. Throughout this paper Dalton talks about *ultimate particles* rather than atoms.

ATOMIC WEIGHTS AND MULTIPLE PROPORTIONS

The first full account of the chemical atomic theory was given in 1807 by Thomas Thomson (1773–1852) in the third edition of his *System of Chemistry*. Thomson had visited Dalton in 1804, and had been converted to Dalton's views on that occasion. Indeed, there is some evidence to suggest that Thomson had been on the verge of a similar theory himself. Dalton published his own account in his book *A New System of Chemical Philosophy*, which appeared in 1808.

By now Dalton was using the term *atom* for the ultimate particles of all substances; elements were composed of simple atoms, and compounds of compound atoms. In order to assign atomic weights, it was necessary to make assumptions concerning the composition of compound atoms. Dalton adopted what he called the *Principle of Simplicity*, in which he stated that, where two elements A and B form only one compound, its compound atom contains one atom of A and one of B. If a second compound exists, its atoms will contain two of A and one of B, and a third will be composed of one of A and two of B, etc. On this basis Dalton proposed that the water atom was composed of one atom of hydrogen joined to one atom of oxygen. The rather poor analytical data available in 1808 indicated that hydrogen and oxygen combined in a weight ratio of 1:7 (rather than 1:8), and thus oxygen was now given a relative weight value of 7 on the basis of hydrogen having a value of 1. Although Dalton's formula for water was wrong, he did suggest the correct atomic composition of many simple compounds, such as carbon monoxide, carbon dioxide, nitrous oxide, nitric oxide and nitrogen dioxide.

Dalton's theory provided a ready explanation for the law of definite proportions. It also predicted that when two elements form more than one compound, the various weights of one element which combine with a fixed weight of the other should be in simple numerical ratio. This prediction, now known as the law of multiple proportions, was found by Dalton to be obeyed by the oxides of nitrogen and by olefiant gas and marsh gas (ethene and methane).

Table 6.1 Dalton's first set of atomic weight values (1805)

Hydrogen	1
Azot	4.2
Carbone	4.3
Ammonia	5.2
Oxygen	5.5
Water	6.5
Phosphorus	7.2
Phosphuretted hydrogen	8.2
Nitrous gas	9.3
Ether	9.6
Gaseous oxide of carbone	9.8
Nitrous oxide	13.7
Sulphur	14.4
Nitric acid	15.2
Sulphuretted hydrogen	15.4
Carbonic acid	15.3
Alcohol	15.1
Sulphureous acid	19.9
Sulphuric acid	25.4
Carburetted hydrogen from stagnant water	6.3
Olefiant gas	5.3

Part and parcel of Dalton's theory was the atomic symbolism that he invented. Earlier chemists and alchemists had used various symbols, but these were little more than a kind of shorthand, although for the alchemists a symbol might have suggested a mystic connection, for example between a metal and a planet. A good example of early chemical symbolism is found in Geoffroy's table of affinities of 1718 (see Figure 4.1). Dalton introduced symbols that had a quantitative as well as a qualitative significance (Figure 6.2). The symbols were circles with a distinctive pattern or letter inside them; each circle represented one atom of that particular element. Compound atoms were represented by touching circles. The arrangement of elementary atoms in the compound atoms was to some extent arbitrary, but Dalton suggested that atoms of the same kind would repel each other.

He thereby arrived at the correct arrangements for carbon dioxide (linear) and sulphur trioxide (triangular).

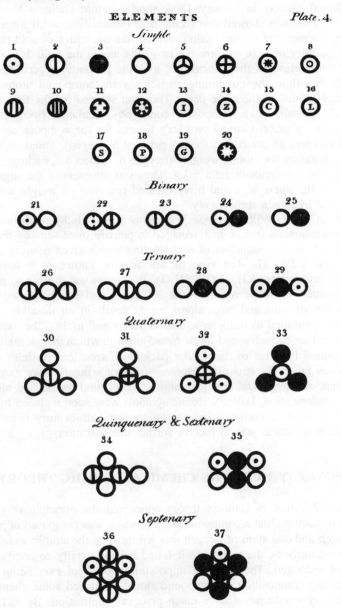

Figure 6.2 Dalton's atomic symbols. Number 28 is a compound atom of carbonic acid (carbon dioxide), and number 31 is a compound atom of sulphuric acid (sulphur trioxide)

As we have seen, it was Thomas Thomson who gave the first published account of Dalton's theory. Thomson also helped to advance it in an important paper read to the Royal Society in January 1808, shortly before Dalton's *New System* was published. Thomson showed that oxalic acid (ethanedioic acid) forms two sets of salts (the normal and the acid salts) and that the quantities of acid that react with a given quantity of base to form the two salts are in the ratio 1:2. This provided another verification of the principle of multiple proportions predicted by Dalton, although this time the combining particles were compound atoms rather than elementary atoms. In the same paper Thomson showed for the first time how the empirical formula of a compound could be calculated from its percentage composition by mass. Using Lavoisier's figures for the composition of sugar (64 per cent oxygen, 28 per cent carbon, 8 per cent hydrogen), Thomson divided them by his own values for atomic weights (oxygen 6, carbon 4.5, hydrogen 1) to obtain figures in the approximate ratio 5:3:4. Thomson represented the sugar particle as 5w + 3c + 4h, where w, c and h represented one relative weight unit of oxygen, carbon, and hydrogen respectively.

Later, in 1808, William Hyde Wollaston (1766–1828) published a paper in which he announced that he had managed to prepare three salts by reacting oxalic acid with a base, the quantities of acid reacting with a given quantity of base being in the ratio 1:2:4. The last type of salt is now known as a tetroxalate (e.g. potassium tetroxalate, $KH_3(C_2O_4)_2.2H_2O$). Wollaston was unable to prepare a salt in which acid and base reacted in a 3:1 ratio, and he speculated that such a combination of acid and base atoms might result in an unstable arrangement. Wollaston continued to think in structural terms and in 1812 he expanded on an idea that Dalton had advanced in the *New System* in which the geometry of crystals was explained in terms of the regular packing of spherical particles (Figure 6.3). This was as far as the structural aspects of Dalton's theory were explored for the time being. Although Dalton's diagrams of compound atoms had also conveyed structural information, Dalton's atomic symbols were soon replaced by a system of letters devised by Berzelius. It was to be another half–century before the spatial arrangement of atoms was to receive further consideration.

CONTROVERSY OVER THE CHEMICAL ATOMIC THEORY

A serious objection to Dalton's theory concerned the principle of simplicity. If Dalton's hypothesis that a compound atom of water was composed of just one atom of hydrogen and one atom of oxygen was wrong, then the atomic weight of oxygen would not simply be the weight of it found experimentally to combine with unit weight of hydrogen. The apparent impossibility of there ever being any way to determine the composition of compound atoms persuaded some chemists that the atomic theory would never have much practical application. By 1814 Wollaston had lost some of his earlier enthusiasm for Dalton's theory, and he proposed that equivalent (or combining) weights be used in place of atomic weights. Richter had calculated equivalents for acids and bases, and Wollaston extended the concept to include salts and individual elements. Wollaston's equivalents were quoted on a scale on which oxygen was assigned a value of 10, and on this basis hydrogen was calculated to have an equivalent of 1.32.

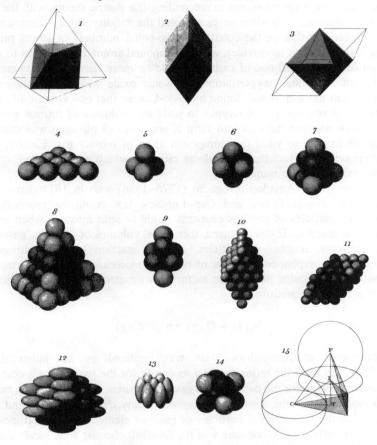

Figure 6.3 Wollaston's suggestions for the packing of atoms in crystals

Wollaston was not rejecting the chemical atomic theory outright, and agreed that it provided the best explanation for the law of multiple proportions. However, he regarded the attempt to calculate atomic weights from combining weights as futile, since it seemed to involve arbitrary assumptions concerning the ratios in which atoms combine. He wrote: 'I have not been desirous of warping my numbers according to an atomic theory'. Wollaston's views had considerable influence, and many chemists used equivalent weights rather than atomic weights.

An important piece of work which appeared to shed some light on the combining ratios of atoms, at least in gaseous reactions, was presented in 1808 by Gay–Lussac. It was already known that hydrogen and oxygen combined in a 2:1 ratio by volume, and Gay–Lussac found that in other reactions simple whole number ratios existed between the volumes of gaseous reactants and products.

These results suggested that equal volumes of gases might contain equal numbers of simple or compound atoms.

However, there were problems in reconciling the atomic theory with the law of combining volumes, and Dalton never accepted the validity of Gay–Lussac's work. Dalton pointed out that on the equal volumes–equal numbers of atoms principle, carbonic oxide (carbon monoxide), whose compound atoms he imagined to contain one atom of oxygen and one of carbon, should be more dense than oxygen itself, consisting of individual oxygen atoms. Carbonic oxide was in fact known to be less dense than oxygen. It was found by Gay–Lussac that one volume of nitrogen combined with one volume of oxygen to yield *two* volumes of nitrous gas (nitric oxide), which was not the expected ratio if one atom of nitrogen was combining with one of oxygen to yield one compound atom of nitrous gas. Finally, Dalton drew attention to the fact that the volume ratios in which gases reacted appeared to be not exactly whole numbers.

It was the Italian Amadeo Avogadro (1776–1856) who in 1811 demonstrated how Dalton's atomic theory and Gay–Lussac's law could be reconciled. He proposed that particles of gaseous elements might be split into two when involved in a chemical reaction. If this occurred, then equal volumes of different gases could contain the same number of particles, and the reaction between nitrogen and oxygen could be explained in terms of two *half molecules* of oxygen combining with two half molecules of nitrogen to form two molecules of nitrous gas; or in terms of a modern equation:

$$N_2(g) + O_2(g) \rightarrow 2NO(g)$$

The acceptance of Avogadro's ideas was hindered by his rather difficult terminology. He used the terms *integral molecule* for the particle of a compound, *constituent molecule* for the particle of a gaseous element, and *elementary molecule* (or *half molecule*) for the atom of an element. Many chemists also found it hard to accept that the fundamental particles of gaseous elements contained more than one atom. An influential voice here was the Swedish chemist Jöns Jacob Berzelius (1779–1848), whose dualistic theory (Chapter 7) precluded the combination of identical atoms. In consequence, Avogadro's hypothesis was poorly received by his contemporaries, and it was not given much attention by the chemical community until 1860, four years after its author had died in relative obscurity.

BERZELIUS AND ATOMIC WEIGHTS

In 1813 Berzelius proposed a system of chemical symbolism which is the basis of the notation used today. He suggested that one volume of an element be represented by the initial letter of its Latin name written as a capital. Where two elements had the same initial letter, the non-metal was assigned the single letter, and a second lower-case letter was added to the symbol of the metal. Although Berzelius opposed Avogadro's concept of the constituent molecule, he became a staunch supporter of Dalton's atomic theory, and his symbols were later used for one atom of an element rather than one volume. Berzelius's symbols were found to be preferable to Dalton's, which were awkward to write and print.

Berzelius (Figure 6.4) studied medicine, but even as a student he was performing chemical experiments in his spare time. After graduation he worked as a physician to the poor in several areas of Stockholm at a very low salary, but continued his chemical investigations in conjunction with Wilhelm Hisinger (1766–1852). They performed important work on the chemical effects of an electric current (Chapter 7). In 1807 Berzelius was appointed professor of medicine and pharmacy at Stockholm, and following a reorganisation in 1810 he became professor of chemistry and pharmacy.

In 1807 Berzelius realised the importance of Richter's concept of equivalents, and he therefore set about obtaining more accurate quantitative analytical data on the composition of salts. Soon after starting this work he learned of Dalton's theory, and he was able to use the data he was acquiring to provide evidence for Dalton's assertions. Over the next ten years Berzelius obtained analytical data on the composition of over 2000 inorganic compounds. He thereby provided extensive verification of the law of multiple proportions, and obtained the data on which he later based his tables of atomic weights. Although he was mainly interested in inorganic chemistry, he also worked on the quantitative analysis of organic compounds (Chapter 8).

Berzelius devised many of the standard laboratory devices and techniques still used today, such as joining glass tubes with rubber connections, using low–ash filter papers, and drying substances in a desiccator.

Another great service that Berzelius rendered to chemistry was the compilation of an annual review of the literature in the physical sciences. The first volume appeared in 1822, and Berzelius continued to compile the sections on chemistry until his death. The *Annual Reports* were translated into German by Wöhler, and thereby became widely read. As a result of compiling the reports, Berzelius was in a position to make some important generalisations. He realised that, in the case of some reactions, the presence of another material was necessary for change to occur, although this substance was not altered in the process. He coined the term *catalysis* for this phenomenon. He also introduced the terms *isomerism, allotropy, polymer* and *protein*.

Berzelius tried to explain all the facts of chemical combination in terms of his dualistic theory (Chapter 7). Later developments in organic chemistry were to cast doubts on dualism, and Berzelius's increasingly desperate attempts to defend the theory meant that towards the end of his life his views were disregarded by other chemists.

Berzelius used superscripts where necessary in the formulae of compound atoms; for example, sulphuric acid (sulphur trioxide) was written as $S\overset{3}{O}$ or SO^3. Later, Berzelius tried to simplify the system by proposing that oxygen in a compound should be indicated by dots above the symbol of the other element. Thus sulphuric acid became

$$\overset{\cdots}{S}$$

He also suggested that two like atoms in a formula should be shown by means of a barred symbol; for example, water could be written as

Ḣ

These modifications were eventually rejected, and our modern system differs little from Berzelius's original concept, except that subscripts (SO_3) are used rather than superscripts.

Figure 6.4 Jöns Jacob Berzelius (1779–1848)

For those who accepted Dalton's theory, the establishment of atomic weights became a matter of prime concern. Oxygen was taken as the standard because most elements form stable and well-defined oxides, although there was no agreement on the value that should be ascribed to oxygen. Berzelius published three tables of atomic weights in the years 1814, 1818 and 1826, and the values in his last table are in most cases in good agreement with the modern ones when recalculated on the basis of O = 16 (Berzelius used the standard O = 100).

To establish the atomic weight of an element, it is necessary to have accurate analytical data for one of its compounds with another element of known atomic weight (e.g. oxygen), and to know the correct formula of the compound. Berzelius was an expert analyst, and he used various methods to deduce the formulae of compounds in preference to putting his faith in Dalton's principle of simplicity.

Because Berzelius could not accept Avogadro's concept of constituent molecules (e.g. N_2) he was unable to accept fully the equal volume–equal number of particles principle of Avogadro, but he did accept that it held for elements which were gaseous. He therefore represented the combination of hydrogen and oxygen as 2H + O, and in this way he was able to assign correct formulae to water and to other compounds formed from gaseous elements.

Two important discoveries were used by Berzelius in revising his atomic weight values before publishing his final table in 1826. The first was the law of isomorphism, which was published in 1820 by the German Eilhard Mitscherlich

(1794–1863). This stated that compounds which crystallise in the same form (i.e. are isomorphous) are similar in atomic composition, and he quoted as examples the sodium phosphates and arsenates, in which an arsenic atom takes the place of a phosphorus atom. Mitscherlich used the fact that potassium selenate was isomorphous with potassium sulphate to deduce the formula of potassium selenate, and hence determine the atomic weight of selenium. Berzelius used the isomorphism of the sulphates and the chromates as evidence that the formula of anhydrous chromic acid was CrO_3. He suggested that the lower oxide of chromium had a formula of Cr_2O_3, and since this was isomorphous with the oxides of iron and aluminium, he was able to propose correct formulae for these compounds.

Berzelius also used the law of Pierre Louis Dulong (1785–1838) and Alexis Thérèse Petit (1791–1820). These two Frenchmen noticed in 1819 that for a number of elements the product of the atomic weight and specific heat (the *atomic heat*) was constant. They had hoped that this would provide an accurate method of determining atomic weights, but the constancy of atomic heats proved to be only approximate. However, Berzelius was able to use this approximate constancy to confirm the formula MO for the oxides of several metals (e.g. Mg, Ca, Ba) in preference to the formula MO_2 which he had used earlier.

PROUT'S HYPOTHESIS

Although Berzelius, in contrast to Dalton, had adopted a rational approach to the determination of the formulae on which his atomic weights depended, many remained sceptical of the chemical atomic theory. Dalton insisted that the atoms of each element were different from the atoms of every other element, and, as the number of known elements increased during the first quarter of the nineteenth century, so the number of postulated individual atoms increased accordingly. Lavoisier listed 31 elements in the *Traité* (excluding light and heat), of which eight were subsequently shown to be compounds. Berzelius's table of 1826 contained 49 elements, and to many people the idea that the universe was composed of such a large number of fundamental particles was absurd. Ever since the philosophers of ancient Greece, those who speculated on the nature of the material world had tended to think in terms of an underlying unity and simplicity, and to many the chemical atomic theory represented a huge step in the wrong direction. Many chemists expected that the so-called 'elements' would ultimately be found to be particularly stable combinations of simpler materials. Typical of this school was Humphry Davy (Chapter 7), who himself decomposed several of Lavoisier's elements, although in each case he discovered a substance which was itself subsequently accepted as an element.

Such was the climate of opinion when, in 1816, William Prout (1785–1850) published his hypothesis that all matter is composed ultimately of hydrogen. Prout cited as evidence the fact that the specific gravities of gaseous elements appeared to be whole–number multiples of the value for hydrogen, and, on the equal volumes–equal numbers of atoms principle, the weights of the atoms of different elements would be whole–number multiples of the weight of the hydrogen atom.

The ancient Greeks had proposed that all matter was composed of a primitive substance called *proto hyle* (Chapter 1), and Prout suggested that proto hyle or *protyle* might in fact be hydrogen.

This idea was enthusiastically adopted by Thomson, who in 1825 published a book entitled *An Attempt to Establish the First Principles of Chemistry by Experiment*, in which he quoted atomic weight values based on determinations made in his own laboratory. For all elements his values were whole–number multiples of the atomic weight of hydrogen. This appeared to provide evidence for Prout's hypothesis, but in fact Thomson had disregarded some small discrepancies from the integral–multiple principle, ascribing the divergencies to experimental error. Berzelius, the leading authority on atomic weight values, was confident that the discrepancies were real, and that Thomson's evidence for Prout's hypothesis was therefore fundamentally flawed. He launched a stinging attack on Thomson's book:

'This work belongs to those few productions from which science will derive no advantage whatever. Much of the experimental part, even of the fundamental experiments, appears to have been made at the writing desk; and the greatest civility which his contemporaries can show its author, is to forget that it was ever published'.

Thomson attempted to defend his position, but the fact that atomic weight values did not conform to a pattern consistent with Prout's hypothesis was soon confirmed by others. Berzelius's attack on Thomson, in that it hinted that the latter had perpetrated some kind of fraud, was entirely unjust. Thomson's values were based on many experiments performed by himself and his students, but he had allowed himself to interpret the results as being consistent with a preconceived notion.

Although this attempt to provide experimental evidence for Prout's hypothesis was a failure, doubts about the chemical atomic theory remained undispelled. Apart from providing a theoretical basis for the laws of definite and multiple proportions, chemical atomism was of little value throughout the first half of the nineteenth century. The continually increasing number of elements troubled many, and the idea persisted that elements were especially stable compound *radicals*, consisting of groups of very simple atoms particularly resistant to decomposition. Berzelius's atomic weights were not widely employed, and many chemists continued to consider reactions in terms of equivalent weights or combining volumes.

After 1860 the chemical atomic theory suddenly acquired great utility. Cannizzaro (Chapter 9) succeeded in convincing the chemical community that the confusion concerning the atomic weights of elements and the formulae of compounds could be removed if Avogadro's hypothesis were accepted. The periodic arrangement of Mendeleev soon followed, and the concepts of valency and structural formulae provided powerful explanations in organic chemistry. Even then, not everyone was convinced, and it was not until early in the twentieth century, when the movement of microscopic pollen grains suspended in water (Brownian motion) was explained in terms of molecular bombardment, that opposition to the atomic theory finally collapsed. By then, different atoms had been found to contain identical subatomic particles, and the first radioactive transmutations had been demonstrated. At that stage, both the upholders of Dalton's theory and those who believed in the fundamental unity of matter could be satisfied. These developments will be considered in later chapters, but it is important to remember that a non-atomic chemistry, or at least a form of atomism very different from Dalton's, seemed to many to be a real possibility throughout the nineteenth century.

7

Electrochemistry and the Dualistic Theory

Lavoisier had defined an element as a substance so far undecomposed. His table of elements in the *Traité* included earths such as lime and magnesia, but he predicted that one day their compound nature would be revealed: '... they will fall to be considered as compounds consisting of simple substances, perhaps metallic, oxydated to a certain degree'. So sure was he that potash and soda were compounds that he omitted them from his list of elements. In 1800 a completely new way of effecting chemical decomposition became available, and with its aid Lavoisier's predictions were verified.

THE VOLTAIC PILE

Before 1800, all electrical investigations had been restricted to static electricity. Electrical charges could be produced by friction and stored in capacitors such as the Leyden jar. The discharge of a capacitor produced a current for such a short time that little progress was made in investigating the chemical effects of an electric current. Priestley and Cavendish had initiated the reaction between hydrogen and oxygen by means of an electric spark, and in 1789 Adrain Paets van Troostwijk (1752–1837) and Jan Rudolph Diemann (1743–1808) decomposed water by means of an electric discharge, but were unable to observe that hydrogen and oxygen were produced at different poles.

The clue needed for the construction of an apparatus which would provide a continuous electric current was provided in 1780 by the Italian Luigi Galvani (1737–1798), who noticed that a freshly dissected frog's leg twitched when in contact with two dissimilar metals. Galvani proposed an explanation in terms of *animal electricity*, but his countryman Alessandro Volta (1745–1827) found that no part of an animal was necessary, and that if two dissimilar metals were in contact with pasteboard soaked in brine, a current would flow in a wire joining the metals. A more pronounced effect was produced by a pile of such units on top of each other (Figure 7.1). Early in 1800, news of the discovery reached England, and within a few days William Nicholson (1753–1815) and Anthony Carlisle (1768–

1840) had constructed a voltaic pile and used it to decompose water. They observed that hydrogen and oxygen were formed at different poles.

As soon as Humphry Davy (1778–1829) heard about this work, he commenced his own research, and before the end of 1800 he had published several papers. He disposed of the view that electricity was generated by the mere contact of two dissimilar metals; he showed that a chemical reaction was taking place. He experimented with various pairs of metals and different solutions, and concluded that one of the metals was being oxidised. He was surprised to find that when he attempted to pass a current through caustic potash solution, the products were the same as those obtained from water.

Davy (Figure 7.2) is one of the most charismatic figures in the history of chemistry. He was born in Penzance, Cornwall, the son of a wood carver. He was undistinguished at school, and on his father's death when he was sixteen he was apprenticed to a local surgeon and apothecary. He suddenly decided that he wished to graduate in medicine, so he embarked on an ambitious plan of self–education. He read Lavoisier's *Traité* and his enthusiasm for chemistry was fired. He started work immediately in his bedroom with apparatus consisting of phials, wine glasses, teacups, tobacco pipes, etc. He was also given an air pump by a friend.

Davy's early experiments concentrated on the nature of heat and light. Lavoisier had listed both as elements in the *Traité*, and thought that gases contained combined heat (caloric). Davy thought he had proved that heat was motion and that only light was matter, and that gaseous oxygen was a compound of elemental oxygen and light. He suggested that it should be renamed *phosoxygen*. Davy wrote an account of his experiments and this was sent to Dr Thomas Beddoes (1760–1808), who was establishing his Pneumatic Institute at Bristol to investigate the medicinal properties of the many recently discovered gases. Beddoes was impressed, and was also very interested to learn that Davy had been experimenting with nitrous oxide. Davy had disproved the theory of contagion of a Dr Mitchill, which ascribed to nitrous oxide terrifying powers of spreading disease. Beddoes offered the nineteen–year–old Davy the post of superintendent of the Pneumatic Institute.

At Bristol, Davy continued to experiment on nitrous oxide and began inhaling it himself, and hence discovered its curious effect on human sensations and behaviour. He also discovered its anaesthetic properties. This work made his name, and his hasty and unfortunate conclusions about phosoxygen, which had been published by Beddoes, were soon forgotten.

After only three years at Bristol, Davy was invited to become Assistant Lecturer in chemistry at the Royal Institution in London. The Royal Institution had been founded by Benjamin Thompson, Count Rumford, who had been born in Massachusetts in 1753, but had fought on the British side in the War of Independence. After the war he came to Europe and entered the service of the elector of Bavaria as Minister for War, during which time he performed his famous cannon–boring experiments (Chapter 13). By 1798 he was in London canvassing the idea of an institution to teach the applications of science and to popularise recent inventions. There was to be a school for mechanics and an exhibition of modern domestic appliances. He obtained sufficient subscribers, and the Royal Institution was founded in 1799.

Soon after arriving at the Royal Institution, Davy delivered a course of lectures on galvanism. The lectures were an outstanding success, and within twelve months Davy had been appointed professor of chemistry. In accordance with the utilitarian aims of the Institution, Davy was asked to perform research on tanning and on agricultural chemistry. Davy incorporated these topics into a general course on chemistry, which he delivered in 1802. Once again, Davy dazzled his audience, and the income that the lectures generated probably saved the Royal Institution from collapse. As a result of Davy's success, the character of the Royal Institution changed; its main public activities became popular lectures, and the plans for the school for mechanics and the exhibition of domestic appliances were abandoned.

Davy was knighted in 1812 and he ceased lecturing at the Royal Institution in the same year, but he continued to work in the laboratory. It has been said that his most important discovery was Faraday, whom he appointed as his assistant in 1813. In 1815, Davy, with Faraday's assistance, devised the lamp named after him for the safe illumination of coal-mines. After this time Davy performed little further significant laboratory work. Davy was not only a brilliant chemist, but he also wrote poetry (he was friendly with Wordsworth, Southey and Coleridge), and he was an expert fisherman.

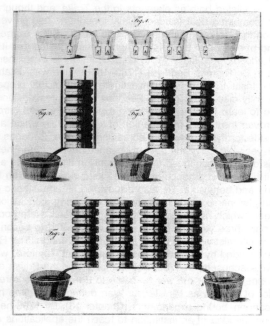

Figure 7.1 Volta's electric piles

Some of Davy's contemporaries thought that the passage of electricity through solutions caused the synthesis of new materials, but he was convinced that decomposition was occurring. The passage of electricity through solutions was also explained in terms of decomposition a few years later by Theodore von Grotthuss (1785–1822). His theory, published in 1805, was that, in the electrolysis of water, decomposition occurred initially near the positive and negative poles, yielding oxygen and hydrogen respectively. The hydrogen and oxygen thus left behind then combined immediately with the appropriate part of a neighbouring particle, thus initiating a concerted chain of decompositions and recombinations throughout the solution. The new particles then turned a somersault, enabling further decompositions to occur (Figure 7.3). Versions of the Grotthuss theory continued to be used until supplanted by the Arrhenius theory in the 1880s (Chapter 13).

Figure 7.2 Humphry Davy (1778–1829)

Davy's move to the Royal Institution meant that he had to abandon his electro–chemical researches for a while. In 1803 Berzelius and Wilhelm Hisinger (1766–1852) found that when an electric current was passed through solutions of various salts, acids were found at the positive pole and bases at the negative. This observation was probably important in leading Berzelius to the dualistic theory (see below).

Davy resumed his electrochemical research around 1806. In a lecture delivered that year, he suggested that chemical affinity was electrical in nature. He saw the passage of an electric current as a means of breaking down compounds that had hitherto resisted decomposition:

'The new mode of analysis may lead us to the discovery of the *true* elements of bodies .... For if chemical union be of the nature which I have ventured to suppose, however strong the natural electrical energies of the elements of the bodies may be, there is every probability of a limit to their strength: whereas the powers of our artificial instruments seem capable of indefinite increase'.

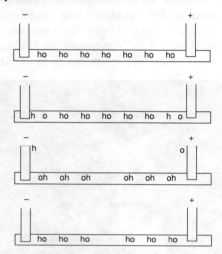

Figure 7.3 A representation of the sequence of events occurring during the passage of an electric current through water (ho) according to the theory of Theodore von Grotthuss, published in 1805

Davy did not develop his theory on the electrical nature of chemical affinity but he did set about trying to decompose potash and soda. As before, passage of electricity through potash and soda solutions yielded only hydrogen and oxygen, but on 6 October 1807 Davy used fused potash and obtained droplets of metallic potassium. Sodium was similarly isolated a few days later (Figure 7.4).

Davy was not sure at first how to classify sodium and potassium, but he decided that, on the basis of their lustre and their conduction of electricity and heat, they should be regarded as metals. He also isolated oxygen from the passage of electricity through fused potash, and since other experiments had led him to the incorrect conclusion that oxygen was present in ammonia, he stated that not only was oxygen an essential component of all acids but also of alkalis as well.

Davy produced only minute amounts of potassium and sodium, and larger quantities were needed for a thorough investigation of their properties. Gay–Lussac and Thenard, working at Arcueil, reasoned that the normal affinities of oxygen for iron and the alkali metals might be reversed at high temperatures. They therefore heated potash and soda with iron filings to a bright red heat in a bent iron gun barrel. The vapour of the alkali metal distilled off, and they were therefore able to prepare reasonable quantities of potassium and sodium.

In 1808 Davy set about trying to decompose earths such as lime and magnesia by the passage of electricity, but ran into difficulties because these materials are infusible. He tried passing the current through a moistened earth mixed with mercuric oxide, and, although some decomposition occurred with the formation of a mercury amalgam, he could not obtain enough of the amalgam to permit isolation of the metal. At this point Davy received a letter from Berzelius reporting that he and M. M. af Pontin (1781–1858) had prepared small quantities of an amalgam by passing a current through lime mixed with mercury. Davy combined the techniques and mixed moist lime with mercuric oxide. He placed the mixture on a platinum plate connected to the positive pole of a pile, and hollowed out a cavity in the mixture into which he poured a globule of mercury, which he connected to the negative pole by a platinum wire. By this method Davy obtained enough calcium amalgam to distil off the mercury and obtain the first sample of calcium. By a similar method Davy quickly isolated barium, strontium and magnesium.

Figure 7.4 Davy's apparatus. On the left is a voltaic pile, the apparatus in the centre is that used by Davy for the isolation of potassium and sodium, and on the right is an apparatus used for the decomposition of solutions

FURTHER ATTEMPTS AT DECOMPOSITIONS

Although Davy had isolated six new metals, he was not certain that they would prove to be elements. Indeed, underlying most of Davy's chemical research was the belief that the number of true elements was probably very small. In his book *Elements of Chemical Philosophy*, published in 1812, Davy stated: 'Matter may ultimately be found to be the same in essence, differing only in the arrangement of its particles, or two or three simple substances may provide all the varieties of compound bodies'.

After his isolation of the six metals, Davy tried to break down other substances that had hitherto resisted attempts at their decomposition. Commenting on the possibility of using potassium and sodium for this purpose he said: 'In themselves they will undoubtedly prove powerful agents for analysis; and having an affinity for oxygen stronger than any other known substances, they may possibly supersede the applications of electricity to some of the undecompounded bodies'. Potassium was in fact used in 1808 to isolate boron from boric acid by Gay–Lussac and Thenard in France, and Davy independently performed the same experiment a few days later.

Davy then made attempts to decompose nitrogen, sulphur and phosphorus. He thought that nitrogen might be a compound of oxygen, and for a time he believed he had detected the presence of traces of both hydrogen and oxygen in sulphur and phosphorus. Davy was toying with the idea of hydrogen being present in all inflammable bodies; he was in effect harking back to something similar to the phlogiston theory.

Davy was unsuccessful in decomposing any more of Lavoisier's elements, but by contrast he showed in 1810 that oxymuriatic acid, which Lavoisier had considered to be a compound and had named accordingly, should be considered to be an element. On Lavoisier's view that all acids contain oxygen, it was generally believed that muriatic acid (hydrochloric acid) was a compound of an unknown element with oxygen, and that the green gas, oxymuriatic acid, obtained by the oxidation of muriatic acid, contained a higher proportion of oxygen. By careful investigation of the reaction of muriatic acid and oxymuriatic acid with potassium, its oxide and its hydrated oxide (hydroxide), Davy concluded that oxymuriatic acid contained no oxygen and was a material rather similar to oxygen. He coined the name *chlorine* on account of its colour (Greek *chloros*, pale green). He also showed that muriatic acid is a compound of chlorine and hydrogen, and thus provided evidence against Lavoisier's theory that oxygen was an essential component of all acids. Once again, similar work was being performed at about the same time by Gay–Lussac and Thenard. Given that chlorine contained no oxygen, and that some substances burned in it, Davy was able to conclude that combustion was not necessarily a combination with oxygen, as Lavoisier had thought.

At about the same time, Gay–Lussac and Thenard were investigating compounds of Lavoisier's *fluoric radical*. Fluoric (hydrofluoric) acid had first been prepared by Scheele, and Gay–Lussac and Thenard obtained the nearly anhydrous acid by distilling fluorspar with concentrated sulphuric acid in a lead retort. Davy also experimented with fluoric compounds. He was in correspondence with André Marie Ampère (1775–1836), who pointed out the similarity between the compounds of chlorine and the fluoric compounds. In 1813, following Ampère's suggestion, Davy coined the name *fluorine* for the element he suspected was present. However, all attempts to isolate the element, either by passing an electric current through hydrofluoric acid, or by reacting oxygen or chlorine with fluoric compounds, met with failure. Davy, Gay–Lussac, Thenard and many others who subsequently worked in this field all damaged their health by breathing fumes of hydrogen fluoride.

In 1813, Davy, accompanied by his newly acquired assistant Michael Faraday (1791–1867), toured France and Italy. In Paris, Davy was given a sample of a violet solid recently obtained from seaweed by Bernard Courtois (1777–1838).

Davy used his travelling case of apparatus and chemicals to show its similarity to chlorine. Yet again Gay–Lussac was conducting a parallel investigation, and he named the material *iode*, which Davy translated as *iodine*. Both men prepared a number of iodine compounds.

Faraday (Figure 7.5) was the son of a blacksmith and he only received an elementary education. At the age of thirteen he was apprenticed to a London bookbinder and took the opportunity of reading some of the books that passed through his hands. One of these was *Conversations on Chemistry* by Jane Marcet (1769–1858), which was written to provide a means by which ladies could educate themselves in chemistry. Mrs Marcet had acquired some of her own knowledge of chemistry by attending Davy's lectures at the Royal Institution. Faraday took every opportunity to extend his knowledge of science, and used some of his meagre resources to buy materials for his own experiments. In 1812 one of the customers at the bookshop gave the enthusiastic Faraday tickets to some of Davy's lectures. Faraday made a fair copy of the notes that he had taken, bound them himself, and sent them to Davy with the request that he be considered for a job. The following year a vacancy arose at the Royal Institution, and Davy offered the position to Faraday.

Faraday is chiefly remembered for his achievements in electromagnetism, but he made some vital contributions to chemistry and indeed he considered that no distinctions should be drawn between the areas in which he worked. Faraday never developed any of his physical ideas mathematically, preferring to put his trust in observation and measurement. He was undoubtedly one of the greatest experimental scientists that has ever lived.

Like Davy, Faraday became a successful lecturer at the Royal Institution. In 1826 he initiated the famous Christmas lectures for young people which still continue, and with the aid of television now reach a very wide audience. He belonged to the strict Sandemanian Christian sect, and when on one occasion he missed attending church on Sunday because he had been commanded to lunch with Queen Victoria, his fellow Elders considered this reason inadequate and he was removed from membership for a while. In 1831 Faraday gave up his lucrative industrial consultancies in order to concentrate on his fundamental research at the Royal Institution. In the 1840s his health started to break down, probably as a result of overwork, and although he continued to lecture until 1861, his most creative period was at an end.

THE DUALISTIC THEORY

Davy had hinted that chemical attraction, or affinity, might be electrical in nature. Berzelius held the same view and developed his dualistic theory, which was in effect an early theory of chemical bonding. According to the theory, atoms carried two charges with either the positive or the negative charge predominating. Metals were classed as electropositive (since they are attracted to the negative pole in electrolysis), and the most electronegative element was oxygen. Chemical combination occurred between elements as a result of attraction between opposite electrical charges. Since atoms carried both charges, it was possible for an atom to be negative towards one element and positive towards another. Thus sulphur was considered to be negative towards metals, but positive towards oxygen.

Figure 7.5 Michael Faraday (1791–1867) lecturing at the Royal Institution

When atoms combined, they did so as a result of electrical attraction, and the resulting compound might still possess an excess of positive or negative charge. Thus copper oxide formed from the union of copper (positive) and oxygen (negative) was slightly positive, and sulphuric acid (SO_3) formed from positive sulphur and negative oxygen was slightly negative. This explained how copper oxide could combine with sulphuric acid to form copper sulphate:

$$CuO + SO_3 \rightarrow CuSO_4$$

Berzelius explained the formation of a hydrate of copper sulphate by saying that copper sulphate was slightly positive, and this enabled it to combine with negative water. Berzelius published his first tentative account of the dualistic theory in 1812.

We can see in the dualistic theory the germ of some modern ideas. Berzelius arranged the elements in order of electro–affinities, oxygen being the most electronegative and potassium the most electropositive. The order is broadly similar to that found in a modern table of electrode potentials. However, Berzelius was reluctant to abandon the central role that oxygen was supposed to play in acids, and his theory would not allow combination to occur between identical atoms. This was one reason why Avogadro's concept of constituent molecules (e.g. N_2) failed to gain acceptance. In inorganic chemistry, the dualistic theory was quite useful, and its echoes are still encountered today (for example, the potassium content of the fertiliser sulphate of potash is often expressed in terms of the percentage of K_2O that it contains).

FARADAY'S CHEMICAL WORK

After his trip to Europe with Davy, Faraday resumed work at the Royal Institution. Initially he assisted Davy, but he gradually increased in confidence and independence. Since it was now known that chlorine would support combustion, Faraday was surprised that carbon, one of the most combustible elements, formed no chlorine compounds. In 1820 he found that when he reacted the compound known as *oil of the Dutch chemists* (1,2–dichloroethane, prepared from the reaction of chlorine with olefiant gas, i.e. ethene) with chlorine in sunlight, he obtained *perchloride of carbon* (C_2Cl_6). This was the first substitution reaction, and reactions of this type were later to present a serious challenge to the dualistic theory. By passing perchloride of carbon through a red–hot tube, Faraday obtained *protochloride of carbon* (C_2Cl_4).

In the 1820s Faraday acquired a growing reputation as a chemist, and he was frequently engaged to perform analytical and consultancy work for various companies. In 1825 he was asked to do an analysis by the Portable Gas Company. This organisation prepared gas for illumination purposes by the destructive distillation of whale and codfish oil. The gas was bottled under pressure and the cylinders were distributed to the Company's customers. Faraday was asked to investigate a liquid that separated from the pressurised gas. By distillation he obtained a liquid that he called *bicarburet of hydrogen*. This was actually the first sample of benzene to be isolated. Faraday also obtained a much more volatile component; this was isobutene (C_4H_8), and Faraday found that it had the same percentage composition as olefiant gas but a vapour density twice as great.

In 1823, Faraday, acting upon a suggestion of Davy, liquefied chlorine by heating chlorine hydrate in a sealed distillation apparatus. Davy then liquefied hydrogen chloride, and Faraday obtained liquid sulphur dioxide, hydrogen sulphide, chlorine dioxide, nitrous oxide, cyanogen and ammonia. These high–pressure distillations were the cause of many explosions from which, fortunately, Faraday suffered no permanent harm. In 1835 Thilorier prepared solid carbon dioxide, and in 1845 Faraday returned to his experiments and cooled the receiver in a mixture of solid carbon dioxide and ether. By this means he liquefied several more gases and solidified hydrogen bromide and hydrogen iodide. However, he failed to liquefy oxygen, nitrogen, hydrogen, methane, carbon monoxide and nitric oxide. Forty years later the liquefaction of gases was again to become an important research topic at the Royal Institution with the work of James Dewar (Chapter 13).

QUANTITATIVE ELECTROCHEMICAL STUDIES

Ever since the discovery of the voltaic pile, there had been doubt about the identity of voltaic and common (i.e. frictional) electricity. In 1831 Faraday made his classic discovery of electromagnetic induction, and thus added a third type of electricity. The first experiment Faraday performed after he had discovered the induced current was to see if it would produce any decomposition in a drop of copper sulphate solution. The quantity of electricity passed was too small to produce any observable effect, and Faraday tried to find a more sensitive chemical test. Using the current from a voltaic pile he decided that touching the wires on a piece of moistened starch–potassium iodide paper and observing the blue colour formed at the point of contact of the positive pole was the best test.

By observing the throw of the needle of a ballistic galvanometer, Faraday estimated that the quantity of electricity flowing when he discharged a set of Leyden jars that had been charged up by 30 turns of a frictional electrical machine was the same as that which flowed in 3.2 seconds from a simple battery he had constructed. Using the iodine test he estimated that the amount of chemical decomposition produced by the passage of these two quantities of electricity was the same. By comparing the colour produced with the deflection of the needle when different quantities of electricity were passed, he was able in 1832 to make his first tentative statement of what later came to be known as his first law of electrolysis: '... It also follows for this case of electrochemical decomposition and it is probable for all cases, that the chemical power, like the magnetic force, is in direct proportion to the quantity of electricity which passes'.

In 1833 Faraday returned to his quantitative work on electrochemical decomposition. Using dilute sulphuric acid, he found that the volume of hydrogen liberated depended only on the quantity of electricity passed and was not affected by the concentration of the acid, the size of the electrodes, or the intensity of the current. He then constructed an apparatus to measure the quantity of electricity by means of the quantity of hydrogen produced by it; such a piece of apparatus he termed the *volta–electrometer*. A few years later this name was changed to the *voltameter*. Today a similar piece of apparatus is called a *coulometer*.

Faraday believed that the passage of an electric current caused a distortion in the forces of affinity which held the compound together. The decomposition was therefore a chemical process, and the quantities of different elements liberated by the passage of the same quantity of electricity should be in proportion to their chemical equivalent weights. In order to test this hypothesis, he passed an electric current through a number of solutions and fused metallic salts, and in each case he also connected a voltameter in series to measure the amount of hydrogen produced by the quantity of electricity that passed. Faraday's predictions were verified, and are now embodied in his second law of electrolysis.

Faraday discussed the theoretical implications of his electrochemical discoveries in 1834:

'... the equivalent weights of bodies are simply those quantities of them which contain equal quantities of electricity, or have naturally equal electric powers; it being the ELECTRICITY which *determines* the equivalent number, *because* it determines the combining force. Or, if we adopt the atomic theory or phraseology, then the atoms of bodies which are equivalents to each other in their ordinary chemical action, have equal quantities of electricity naturally associated with them. But I must confess I am jealous of the term *atom*; for though it is very easy to talk of atoms, it is very difficult to form a clear idea of their nature, especially when compound bodies are under consideration.'

Thus, although Faraday disliked thinking in terms of atoms, he clearly saw the possibility of unitary atomic electrical charges.

The new science of electrochemistry to which Faraday had made such important contributions needed a new terminology. Faraday sought the advice of the classical scholar William Whewell (1794–1866), who suggested the terms *electrode, anode, cathode, ion, anion, cation, electrolyte* and *electrolysis*. The concept of charged particles was still a long way in the future; the term *ion* meant traveller, and thus an *anion* was a substance that travelled to the anode.

DOUBTS ABOUT THE CHEMICAL ATOMIC THEORY

We have seen (Chapter 6) that Dalton's theory received a somewhat cool reception. Although Berzelius was a great champion of the theory, many others remained unconvinced, and among these were Davy and Faraday. The chief stumbling blocks were Dalton's arbitrary assumptions about the formulae of compounds, and the large number of different atoms that the theory seemed to require. Dalton's rejection of Gay–Lussac's law of combining volumes also seemed irrational to many. In this climate, many chemists considered Dalton's important contribution to be the law of multiple proportions, and they used Wollaston's equivalent weights rather than the atomic weights of Dalton or Berzelius.

There was another kind of atomism, which probably appealed to Davy and certainly appealed to Faraday. This was derived from the theory originally proposed in 1758 by the Jesuit priest Rudjer Boscovic (1711–1787). This considered that atoms were not tiny material particles, but point centres of force. When they were far apart, two point atoms attracted each other according to an inverse square law. As the distance between the atoms decreased, the force switched from attraction to repulsion several times. As the distance between the point atoms approached zero, the repulsive force tended to an infinite value. This complex pattern of forces made it reasonable to suppose that the point atoms could form many different stable aggregates, and these different aggregates represented the particles of the chemical elements. Towards the end of his life Faraday wrote: 'To discover a new element is a very fine thing, but if you could decompose an element and tell us what it is made of—that would be a discovery indeed worth making'.

During the first half of the nineteenth century, it was the discoveries in organic chemistry that were to have the greatest implications both for electrochemical dualism and for the chemical atomic theory. Although Berzelius defended the dualistic theory until his death, the inability of the theory to provide satisfactory explanations for the discoveries of the 1830s and 1840s led to its eventual abandonment. The rapidly growing number of carbon compounds also presented difficulties for the atomic theory, as uncertainties in atomic weight values meant that no agreement was possible on the formulae of the new compounds that were being produced.

8

The Foundation of Organic Chemistry

In the first half of the nineteenth century, many chemists were reluctant to think in terms of atoms of uncertain weight and molecules of uncertain formula. But for those who wished to utilise the atomic theory, the greatest challenge came from the field of organic chemistry. It was as a result of growth in this area during the first 60 years of the nineteenth century that it became imperative to settle the question of uncertain atomic weights and to determine accurate molecular formulae. The attempts to systematise the growing number of carbon compounds also led, in the 1850s, to the concept of valency. In this chapter we shall be concerned with the development of organic chemistry up to the emergence of this idea, one of the most fundamental in the science.

ORGANIC SUBSTANCES

Materials obtained from animal and vegetable sources had been classified and studied separately from those obtained from mineral sources ever since the days of the Arab alchemists. Bergman in 1790 first referred to 'inorganic and organic bodies', and Berzelius in 1806 first used the term 'organic chemistry'. As knowledge concerning organic bodies increased, and more compounds that do not occur in nature were prepared from them, the term *organic chemistry* gradually became synonymous with the chemistry of carbon compounds.

The alchemists and early chemists often worked with materials such as blood, saliva, urine, egg–white, gum, etc. One of the first organic compounds obtained in a fairly pure state was alcohol (ethanol). It was prepared by distillation in Europe in the twelfth century, and it may also have been familiar to the Arab alchemists. Ethanol had also been obtained many centuries earlier in China, both by distillation and by removing the water from a dilute solution by freezing. In the sixteenth century, the products of the reaction of alcohol with each of the three mineral acids had been prepared. Before 1780 only four organic acids were known: formic, acetic, benzoic, and succinic. Around this time, Scheele isolated several more, including oxalic, tartaric, citric and lactic. He also obtained a number of esters, and he isolated glycerol and hydrocyanic acid.

After Scheele's death in 1786 there was something of a respite in the discovery of new compounds from natural sources, but the first three decades of the nineteenth century saw a huge increase in the number of organic compounds known in the pure state. Among these were the three sugars glucose, fructose and sucrose, isolated by Proust, who also obtained the amino acid leucine from cheese. It was demonstrated by Gottlieb Sigismund Kirchhoff (1764–1833) that starch yielded glucose when boiled with sulphuric acid. By 1835 many alkaloids had been isolated including morphine, strychnine, quinine and caffeine. The French chemist Michel Eugène Chevreul (1786–1889) demonstrated that fats are compounds of glycerol and fatty acids, several of which he isolated for the first time. In the 1820s he proposed that a sharp melting temperature was a criterion of purity for an organic compound. Chevreul is the only major figure in the history of chemistry who lived to be over 100, and he played an active role in French science until his death.

The increasing number of carbon compounds being discovered led to attempts to rationalise and understand the mass of facts they presented. It was chemists in France and Germany who were most prominent in this task, although important contributions were also made by a small number in Great Britain and by the Swede Berzelius.

VITALISM

Lavoisier, when considering organic compounds, had attached no special significance to the fact that they had been made by living organisms. However, in the early years of the nineteenth century, many chemists adopted a contrary view, namely that organic compounds could only be produced through the agency of a 'vital force' which was present in living plants and animals. One of the enduring myths in the history of chemistry is that vitalism was disproved in 1828 when Friedrich Wöhler (1800–1882) made urea. Wöhler was attempting to prepare ammonium cyanate by reacting silver cyanate with ammonium chloride. The ammonium cyanate had been converted to urea while the solution was being evaporated to dryness:

$$NH_4CNO \rightarrow NH_2CONH_2$$

Wöhler wrote to Berzelius: 'I can make urea without the necessity of kidney, or even an animal, whether man or dog'. However, he made no claim to have disproved vitalism, and both the ammonia and the cyanic acid that he used had been derived from natural sources. The importance of Wöhler's urea synthesis was that both ammonium cyanate and urea had the same composition, and it therefore provided an early example of isomerism.

Wöhler (Figure 8.1) was the son of a veterinary surgeon. He studied medicine but decided to devote his life to chemistry. After graduating in 1823, he spent one year working with Berzelius in Sweden and then returned to Germany. His first post was as a teacher in a technical school in Berlin, and his laboratory was in the room where a fraudulent alchemist called Count Ruggiero had worked before he was hanged on a gilded scaffold in 1709. In 1831 Wöhler moved to Castile and in 1836 he became professor of chemistry at the University of Göttingen, where he remained for 21 years. He isolated metallic aluminium in 1827, and although Oersted had obtained the element two years earlier, Wöhler's sample was the first that was sufficiently pure to enable its properties to be determined. Wöhler also studied the chemistry of beryllium, boron, silicon, titanium, vanadium and yttrium. Wöhler's teaching laboratory at Göttingen became almost as attractive to aspiring chemists as that of Liebig at Giessen, and probably his most famous pupil was Hermann Kolbe (1818–1884).

Figure 8.1 Friedrich Wöhler (1800–1882)

Vitalism did receive a serious setback in 1844 when Hermann Kolbe (1818–1884) synthesised acetic acid from non–organic materials. It was already known that iron pyrites and carbon when heated together produce carbon disulphide. Kolbe showed that treatment of carbon disulphide with chlorine resulted in the production of tetrachloromethane, which could then be converted to tetrachloroethene by passing the vapour through a red–hot tube. He also found that tetrachloroethene reacted with chlorine and water in the presence of sunlight to produce trichloroacetic acid. Since Louis Henri Frédéric Melsens (1814–1886) had already shown that trichloroacetic acid could be converted to acetic acid, Kolbe could state '... acetic acid, hitherto known only as an oxidation product of organic matter, can be almost immediately composed by synthesis from its elements'. The doctrine of vitalism was then gradually abandoned, but was still thought to be applicable to biochemical reactions such as fermentation until the end of the nineteenth century.

ORGANIC ANALYSIS AND ORGANIC FORMULAE

The large increase in the number of organic substances being reported in the early part of the nineteenth century was accompanied by refinements in the techniques of quantitative elemental analysis used to establish the composition of the new compounds. The original method of analysis for carbon and hydrogen had been developed by Lavoisier, who burned the organic compound in air or oxygen and collected and weighed the carbon dioxide and water produced. However, he found the problems associated with the method considerable. Referring in the *Traité* to the combustion of oils, he states that the difficulties he experienced 'are so insurmountable and troublesome that I have not hitherto been able to obtain any rigorous determination of the quantities of the products'. Gay–Lussac and Thenard made a significant improvement to the technique in 1810 when they reacted the organic compound with the oxidising agent potassium chlorate. In 1815 they replaced this with copper oxide. Further refinements were introduced by Berzelius and in 1831 by Justus von Liebig (1803–1873).

Liebig (Figure 8.2) was born in Darmstadt where his father was a pharmacist. He was apprenticed to another pharmacist but the apprenticeship was terminated when Liebig's father could not pay the agreed indenture fee. It was once thought that Liebig had been dismissed because his unauthorised chemical experiments resulted in an explosion, but this story is now known to be the invention of one of his biographers. Liebig then spent two years helping with the family business, after which time the family fortunes had improved and he was sent to the University of Bonn to study chemistry. He completed his education in Paris under Gay–Lussac. He returned to his native Germany to take up the position of professor of chemistry at the small University of Giessen. Not only did he perform important research, but also at Giessen he established the first proper teaching laboratory. Previously, a professor would allow the occasional gifted student to work in his laboratory, but under Liebig's leadership the number of students trained in the methods of chemical research was much greater, and Giessen was soon attracting students not only from Germany but also from other countries as well. Among those who worked in Liebig's laboratory were the German August Wilhelm von Hofmann, the Frenchmen Charles Gerhardt and Charles Adolph Wurtz, and the Englishmen Alexander William Williamson and Edward Frankland. All five were to make important contributions to the development of organic chemistry. Other universities, especially in Germany, soon opened their own teaching laboratories, often with a former student of Liebig or Wöhler in charge. In consequence, the number of German chemical researchers increased considerably, and the German contribution to chemistry, especially organic chemistry, increased accordingly. Furthermore, the number of qualified chemists being produced enabled Germany to establish a chemical industry that became the most powerful in the world.

As well as perfecting the technique for estimating the carbon and hydrogen content of compounds (Figure 8.3), Liebig also devised quantitative analytical methods for sulphur and the halogens, and Jean Baptiste André Dumas (1800–1884) was responsible for introducing an accurate method of analysis for nitrogen.

The oxygen content of an organic compound was estimated by the difference between the total percentages of all the other elements present and 100 per cent. But although accurate elemental analysis became routine, the deduction of the formula of a compound from the analytical data was still fraught with difficulty. Today, given that the composition of acetic (ethanoic) acid is C 40.0 per cent, H 6.7 per cent, O 53.3 per cent, it is a relatively simple matter to deduce that the empirical formula of the compound is CH_2O. Such a calculation employs modern atomic weight (relative atomic mass) values of C = 12, H = 1, O = 16; but many chemists in the period 1830–1860, for example Kolbe, used Wollaston's equivalents (C = 6, H = 1, O = 8) and hence would calculate the empirical formula to be CHO. There then remained the problem of how many empirical formula units were contained in the molecular formula of the compound. Many compounds were given formulae which, compared with their modern counterparts, contained double the number of atoms of each element.

Figure 8.2 Justus von Liebig (1803–1873)

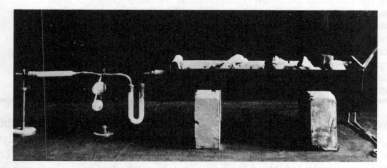

Figure 8.3 Liebig's apparatus for the elementary analysis of organic compounds

Dumas (Figure 8.4) was born in the small French town of Alais where he was apprenticed to an apothecary at the age of 15. However, he wished to obtain a proper scientific education, so he went to Geneva to attend lectures on chemistry, physics and botany, and in 1823 he moved to Paris to complete his chemical studies. Within a few years he had become the most prominent chemist in the country, and a leading member of the French Academy of Sciences. Like Liebig, he established a teaching laboratory, but he had to finance it himself. After the political upheavals of 1848 his income from his teaching posts was reduced and he was forced to close the laboratory. He then played an important part in public affairs and became Minister of Education.

Figure 8.4 Jean Baptiste André Dumas (1800–1884)

The solution to these problems was to come from vapour–density measurements, but, although the appropriate experimental technique was developed by Dumas in 1826, it was to be another 34 years before the confusion was resolved. It will be recalled that in 1811 Avogadro had proposed that equal volumes of gases contain equal numbers of particles. This enabled atomic weights of gaseous elements to be determined from their densities (e.g. the density of oxygen is 16 times that of hydrogen). However, at the time, only four gaseous elements were known: hydrogen, oxygen, nitrogen and chlorine. Dumas devised a technique by which the density of the vapour from a volatile liquid or solid element could be determined. Dumas used this technique to determine the atomic weights of iodine, sulphur, mercury and phosphorus, and Mitscherlich made measurements on bromine and arsenic. Following Avogadro, they assumed that all elements exist as diatomic molecules in the gaseous phase. The values they obtained for the atomic weights of bromine and iodine accorded with those obtained by other methods, but in the other cases the results were anomalous. We now know that mercury in the vapour phase is monatomic, that phosphorus and arsenic are tetratomic, and that at the temperature of Dumas's experiments sulphur vapour consists of a range of molecular sizes.

The immediate effect of these experiments was to cast further doubt on Avogadro's hypothesis. However, Dumas had also used his technique to measure the vapour density of some volatile compounds. It was clear to the pioneers of organic chemistry that, if the equal volumes–equal numbers of particles principle of Avogadro was valid, then the vapour density of an organic compound would be proportional to its molecular weight. Those who did accept this principle obtained molecular weight values in the correct ratio but frequently all the values were double those accepted today. The formulae derived from such values were called *four–volume* formulae, as the weight of four volumes of the organic vapour was compared with the weight of one volume of hydrogen. This system was adopted by Liebig, and the molecular weights it generated were consistent with those adopted by Berzelius for organic compounds.

As well as the problems concerning organic formulae, there were instances of two compounds possessing the same elemental composition and yet displaying different properties. The first example of this phenomenon was discovered in 1823. Wöhler prepared silver cyanate (AgCNO), while Liebig, working in Gay–Lussac's laboratory in Paris, prepared silver fulminate (AgOCN). The two compounds were found to give the same results in quantitative elemental analysis, and yet were clearly different. In 1825 Faraday isolated a hydrocarbon (butene) with the same empirical composition of ethene but twice its density. In 1828 Wöhler converted ammonium cyanate into urea, but both compounds had the same composition. In 1830, Berzelius, who was still the leading chemical authority in Europe, abandoned the position that different properties must be the result of different composition, when he became convinced that racemic and tartaric acids were different compounds that yielded the same results when analysed. Berzelius suggested the name *isomerism* for the new phenomenon from two Greek words meaning *equal parts*. He subsequently described two types of isomerism; the term *metamerism* was applied to cases in which the molecular weights of the compounds were the same, and *polymerism* to cases where the molecular weight of one compound was found to be an integral multiple of the other. Today the term 'isomerism' covers only those cases described by Berzelius under metamerism.

THE RADICAL THEORY

The previously noted confusion over formulae was to continue until after 1860, but it did not prevent the leading chemists of the day from developing theories concerning the nature of organic compounds. It will be recalled that de Morveau and Lavoisier had introduced the concept of the radical in 1787. When a radical combined with oxygen, an acid was formed. For mineral acids, the radical was an element, whereas for acids derived from substances in the animal or vegetable kingdoms, the radical was invariably a compound containing both carbon and hydrogen. The radicals of different organic acids contained carbon and hydrogen in different proportions.

In 1815 Berzelius published his analyses of a number of lead salts of organic acids. He represented organic salts, like inorganic salts, in dualistic terms. Thus lead sulphate was $SO_3.PbO$ (sulphuric acid being represented as SO_3, which we now regard as the anhydride of sulphuric acid), and Berzelius gave lead acetate the formula $C_4H_6O_3.PbO$, thus assigning acetic acid the formula $C_4H_6O_3$.

The concept of the radical was extended in 1815 by Gay–Lussac, who investigated cyanogen (C_2N_2), hydrogen cyanide and the salts of hydrocyanic acid. He noticed that the CN group persisted through a sequence of reactions, with properties somewhat similar to chlorine or iodine. The CN group was called the *cyanide radical*, and the term radical thus acquired the meaning of a particularly stable group of atoms. The first such group in organic chemistry was proposed in 1828 by Dumas and Polydore Boullay. They suggested that olefiant gas (ethene) possessed an alkaline character, and just as one volume of ammonia reacted with one volume of hydrogen chloride, so one volume of olefiant gas reacted with one volume of hydrogen chloride to produce hydrochloric ether (chloroethane). A number of other compounds could be formulated to demonstrate the presence of olefiant gas.

Although the term 'radical' was not used by Dumas, it was employed by Liebig and Wöhler in the report of a famous piece of research published in 1832. They isolated oil of bitter almonds (benzaldehyde) and converted it to several other compounds including benzoyl chloride and benzoic acid. Analysis of all the compounds revealed the presence of a common group, which they called the *benzoyl radical* and formulated as $C_{14}H_{10}O_2$ (actually C_7H_5O).

Berzelius was initially enthusiastic about the discovery of a radical that contained three elements, but on further reflection he decided he could not accept the presence of oxygen in a radical. He maintained that organic radicals should contain carbon and hydrogen only, and their properties would be modified by combination with oxygen. He proposed that some hydrocarbon radicals on combination with oxygen yielded acids such as acetic acid ($C_4H_6O_3$) while others produced substances akin to metallic oxides. Typical of the latter was ethyl oxide (ether), $C_4H_{10}O$. Such compounds would be expected to combine with acids to form 'salts', for example $C_4H_6O_3.C_4H_{10}O$. This compound is ethyl acetate (double formula) and thus Berzelius was able to extend his dualistic ideas into a new area of organic chemistry. The difficulty that it is alcohol rather than ether that combines with acetic acid to produce the ester was overcome when Liebig suggested that alcohol is the hydrate of ether ($C_4H_{10}O.H_2O$).

SUBSTITUTION AND THE TYPE THEORY

Soon after Liebig and Wöhler published their work on benzoyl compounds, which was largely responsible for the enthusiasm with which the radical theory was adopted, Dumas commenced a study of substitution reactions. The results of these researches were to modify the concept of the radical and to undermine Berzelius's ideas on the application of dualism in organic chemistry.

Dumas's research was prompted by an incident during a reception given by Charles X in the Tuileries in Paris. The candles burned with a smoky flame and emitted choking fumes. Alexander Broignart, the director of the Sèvres porcelain factory, was asked to find the cause, and he referred the problem to Dumas, who was his son–in–law. Dumas established that the fumes were due to hydrogen chloride derived from the chlorine that had been used to bleach the wax. He found that the chlorine was not present in the wax as an impurity, but had become chemically combined with it. He studied the reaction of chlorine, bromine and iodine with oil of turpentine and other substances, and concluded that the halogen was removing hydrogen and replacing it, and that for every volume of hydrogen lost from the compound, an equal volume of halogen entered it. Dumas pointed out that other cases of substitution had been reported previously by Faraday and Gay–Lussac, and by Liebig and Wöhler when they converted benzaldehyde to benzoyl chloride.

The study of substitution was then taken up by August Laurent (1808–1853), a pupil of Dumas. He drew attention to the fact that the properties of the chlorinated product did not seem to be appreciably different from those of the starting material. It appeared that the chlorine was exerting an effect similar to that of the hydrogen which it had replaced. Berzelius poured scorn on this idea, saying that it was impossible for an electropositive hydrocarbon radical to contain chlorine and retain similar properties. Electronegative chlorine, according to his dualistic theory, was incorporated into compounds by attraction to an electropositive entity. Dumas wilted under Berzelius's attack, and while he stood by his experimental observations, he distanced himself from Laurent's theoretical views. Laurent was undeterred and further developed his theory, maintaining that there exist both fundamental and derived radicals. A fundamental radical contained only carbon and hydrogen, and was converted into a derived radical with generally similar properties by substituting one or more of the hydrogen atoms by those of another element. This was a complete departure from the concept of a radical as an unchangeable group of atoms, and Laurent eventually replaced the term radical with that of *nucleus*.

Dumas remained unconvinced until 1838, when he prepared a chlorinated acetic acid (actually trichloroacetic acid) and noted its similarity to acetic acid. He then advanced his own version of Laurent's theory, calling it the *theory of types*: 'In organic chemistry there exist certain Types which persist even when in place of hydrogen they contain an equal volume of chlorine, bromine or iodine'. Laurent complained bitterly that Dumas had borrowed his ideas without due acknowledgement, and a rift arose between the two men which prevented Laurent's advancement within the French chemical establishment. Dumas proceeded to make some wild and unsubstantiated claims for the type theory, suggesting that, within a type, other atoms, including those of carbon, might be substituted. This proposal was greeted with derision by Wöhler and Liebig. The former, writing in the journal *Annalen der Chemie* under the pseudonym S. C. H. Windler (=Swindler), reported that he had replaced all the atoms in manganous acetate by those of chlorine without affecting its properties. Liebig, the editor of the journal, added a footnote saying that bleached fabrics consisting entirely of chlorine were on sale in London, and were very popular.

DUALISM IN RETREAT

In spite of the theoretical excesses of Dumas, the experimental facts of substitution were slowly eroding the position of Berzelius, who still clung tenaciously to his dualistic theory. Berzelius was frequently forced to invent new sets of radicals so that he could continue to write formulae in dualistic terms that were consistent with the new discoveries. Thus he represented trichloroacetic acid as a combination of the radicals C_2Cl_6 and C_2O_3, which then combined with H_2O. However, the parent acetic acid was represented as C_4H_6 plus O_3 combined with water. The idea that replacement of hydrogen by chlorine should cause such a fundamental rearrangement of radicals was rejected by many, and it became increasingly common to write unitary formulae, e.g. $C_4H_8O_4$ for acetic acid and $C_4H_2Cl_6O_4$ for trichloroacetic acid. With such a representation it was easy to see that both compounds belonged to the same chemical type.

In 1839 another Frenchman, Charles Frédérick Gerhardt (1816–1856), introduced his theory of residues, which stated that organic reactions could be considered to be the separation and rejoining of *residues*, which he regarded as being somewhat similar to radicals. This type of reaction he at first called *copulation*, but later he used the term *double decomposition*. Thus the nitration of benzene, a reaction first reported by Mitscherlich in 1833, was regarded as the elimination of part of each reactant to form water and the combination of the remaining parts (residues) to form nitrobenzene. In terms of a modern equation this would be represented as:

$$C_6H_5H + HONO_2 \rightarrow C_6H_5NO_2 + H_2O$$

Gerhardt did not believe the structure of a molecule was capable of elucidation, and he wrote the formula of the same compound in terms of different residues in order to explain different reactions.

The credibility of Berzelius's dualistic theory as applied to organic compounds received a further blow in 1842 when Dumas's assistant Melsens reported that trichloroacetic acid could be converted to acetic acid by the action of electrolytic hydrogen, the chlorine being simply replaced by hydrogen. Even Berzelius was now forced to admit that acetic acid and trichloroacetic acid should be formulated similarly, so he now proposed that both acids consisted of a radical combined with oxalic acid (C_2O_3). Thus:

acetic acid $C_2H_6 + C_2O_3 + H_2O$
trichloroacetic acid $C_2Cl_6 + C_2O_3 + H_2O$

Berzelius had thus tacitly accepted the fact of substitution and had attempted to incorporate it into his dualistic theory. He applied Gerhardt's terminology to radicals such as C_2H_6 and C_2Cl_6, calling them *copulae* and referring to the compounds as *copulated*. This new stance by Berzelius won little support, and the dualistic theory was further undermined by an accumulating mass of evidence on the composition of acids.

Lavoisier's idea that acids were composed of radicals and oxygen was one of the cornerstones of Berzelius's dualism. By 1815 both Davy and Gay–Lussac had suggested that hydrogen plays an essential role in acids as a result of their work on muriatic (hydrochloric) and hydrocyanic acids respectively. In 1838 Liebig resurrected these ideas and proposed that acids are hydrogen compounds in which the hydrogen may be replaced by metals. He stated that the hydrogen is not present as water and is located outside the radical. The water formed when an acid reacts with a metal oxide to produce a salt could arise by the combination of the hydrogen of the acid with the oxygen of the metal oxide.

In 1843 Gerhardt suggested that many organic formulae then employed might be double the correct values. He reached this conclusion on the grounds that when simple substances such as water, ammonia or hydrogen chloride are eliminated in organic reactions, they are always produced in pairs, assuming the correctness of Berzelius's formulae of H_2O, NH_3 and HCl for these compounds. Thus Gerhardt proposed that the current organic formulae be halved, and on this basis trichloroacetic acid became $C_2HCl_3O_2$, and could not contain a molecule of water in the formula as Berzelius maintained. Likewise, silver acetate became $C_2AgH_3O_2$, and could not contain a molecule of silver oxide, which Gerhardt suggested had the formula Ag_2O. Gerhardt summarised his views by saying 'there is no water in our acids and no oxide in our salts'.

The new formulae of Gerhardt were called *two–volume* formulae, because the molecular weight of a volatile compound was the weight of two volumes of its vapour compared to the weight of one volume of hydrogen. Gerhardt was therefore ascribing an atomic weight of 12 to carbon and 16 to oxygen. These were the values advocated by Berzelius. Confusion over organic formulae was not terminated with the publication of Gerhardt's paper. Gerhardt himself used the old formulae in his textbook of organic chemistry because he thought that this would attract purchasers.

FREE RADICALS

Although the dualistic theory was in retreat, the concept of the radical was still alive. An important advance for the radical theory came with the work performed on organic arsenic compounds by the German Robert Wilhelm Bunsen (1811–1899) between 1837 and 1843. He started with the dense, evil–smelling, spontaneously inflammable liquid called *cacodyl*, which the French pharmacist Cadet–

Gassicourt had obtained in 1760 by distilling potassium acetate with arsenious oxide. From this liquid Bunsen isolated cacodyl oxide ($C_4H_{12}As_2O$). Hydrochloric acid converted the oxide to the chloride, and when this was treated with zinc Bunsen obtained what he considered to be the free cacodyl radical [$C_4H_{12}As_2$; now known to be $(CH_3)_2As.As(CH_3)_2$]. It had always been supposed that organic radicals should be capable of independent existence, and this apparent isolation of a free radical gave a considerable boost to the theory. Bunsen terminated his researches on cacodyl compounds in 1843 when an explosion of cacodyl cyanide resulted in the loss of his right eye. He subsequently developed the gas burner named after him (surely the best-known piece of chemical laboratory equipment), and was involved in the development of several other pieces of apparatus including the spectroscope (Chapter 9).

Bunsen performed most of his work on cacodyl compounds while professor at Marburg, where Hermann Kolbe started work in 1842 as Bunsen's assistant. Kolbe had been a student of Wöhler and became a firm supporter of the radical theory. He continued to apply Berzelius's dualistic ideas to organic compounds at a time when they had been abandoned by most other chemists. In 1847, after his synthesis of acetic acid from non-organic materials, Kolbe went to London to work as assistant to Lyon Playfair (1818–1898). He soon formed a friendship with Edward Frankland (1825–1899), who had come to London to study with Playfair after serving an apprenticeship with a pharmacist in his native Lancashire. Kolbe and Frankland, working together first in London and then in Bunsen's laboratory in Marburg, discovered that organic cyanides (nitriles) could be hydrolysed to carboxylic acids. This reaction is important today in organic synthesis, but at the time it appeared to provide new evidence in favour of dualism. Berzelius had formulated acetic acid as containing the oxalate radical (C_2O_3), and since cyanogen itself on hydrolysis gives oxalic acid, it appeared that all fatty acids could be regarded as oxalic acid copulated to an alcohol (alkyl) radical.

Frankland returned to London while Kolbe continued to work in Germany on the nature of fatty acids. Kolbe reasoned that if the dualistic formulation of the acids were correct, they might be susceptible to electrolytic decomposition. He found that aqueous potassium acetate was decomposed by an electric current, but not quite in the way he anticipated. On the basis of Berzelius's formula for acetate, he expected that the electronegative oxalate radical would be discharged at the positive pole, where it would combine with oxygen from the electrolysis of water to form carbon dioxide. At the negative pole he expected the evolution of methane, from the combination of hydrogen and the methyl radical. Carbon dioxide was indeed evolved at the positive pole, but at the negative pole he obtained hydrogen and a new organic gas that he took to be the free methyl radical (actually ethane).

Meanwhile Frankland had been pursuing his own attempts to isolate free radicals. His method was to react together a metal with an organic cyanide or halide in the hope that combination would occur between the metal and the halide or cyanide, leaving the organic radical in the free state. The most important experiments involved zinc and ethyl iodide (iodoethane). The reactions occurring in terms of modern equations are:

$$2C_2H_5I + Zn \rightarrow C_4H_{10} + ZnI_2$$
$$2C_2H_5I + 2Zn \rightarrow (C_2H_5)_2Zn + ZnI_2$$

Frankland thought the butane produced was the free ethyl radical (which he formulated as C_4H_5 on the basis $C = 6$). Corresponding results were obtained when zinc was reacted with iodomethane. Not only did these experiments appear to result in the production of free radicals, but also they opened up the field of organometallic chemistry. Organometallic compounds were to have wide application as synthetic intermediates (Chapter 10).

Frankland soon prepared more organometallics by reacting other metals with halogenoalkanes. The reaction mixtures were confined in sealed tubes and heated by focusing the sun's rays upon them. Frankland prepared the methyl and ethyl derivatives of mercury, tin and antimony, and also such compounds as 'oxide of stanethylium' (diethyltin oxide, $(C_2H_5)_2SnO$), which he formulated as C_4H_5SnO.

Kolbe had recently suggested that the cacodyl radical was composed of two methyl radicals *conjugated* to arsenic, and that the cacodyl compounds discovered by Bunsen (e.g. cacodyl oxide and cacodyl chloride) were analogous to the corresponding chloride and oxide formed by arsenic alone. Kolbe was implying that when an element is conjugated to an organic radical its chemical character remained unaltered. When Frankland surveyed the range of compounds he had prepared, he found it impossible to support Kolbe's views. While 'oxide of stanethylium' might be analogous to SnO, it was impossible to prepare a higher oxide of stanethylium analogous to SnO_2. He therefore suggested that atoms have certain fixed combining powers which were independent of the nature of the atoms to which they were united. He cited as evidence the elements nitrogen, phosphorus, antimony and arsenic, which always seemed to combine with three or five atoms of another element.

THE NEW TYPE THEORY

While Frankland was evolving his theory of combining power, the *new type* theory was making its appearance. The primary amines methylamine and ethylamine had been prepared in 1849 by Charles Adolphe Wurtz (1817–1884), and he recognised that these compounds were related to ammonia. The work was continued by August Wilhelm von Hofmann (1818–1892), who in 1851 prepared primary, secondary and tertiary amines and also quaternary ammonium salts. He classified these as belonging to the *ammonia type* (Figure 8.5).

At the same time Alexander William Williamson (1824–1904) was trying to prepare higher alcohols by substituting the hydrogen in ethanol by an alkyl radical. When he reacted ethanol with potassium ethoxide, he found that instead of another alcohol he obtained diethyl ether. This synthesis meant that ethanol could not be the hydrate of ether as Liebig had proposed, and Williamson suggested the existence of a *water type*.

Williamson proposed that acetic acid be regarded as belonging to the water type, with one hydrogen atom replaced by C_2H_3O. He also predicted the existence of a substance in which both hydrogen atoms of water are replaced by C_2H_3O, and this compound (acetic anhydride) was prepared by Gerhardt in 1852. In 1853 Gerhardt added two further types: the hydrogen type, which included the hydrocarbons; and the hydrogen chloride type, which included the halogenoalkanes.

$$\left.\begin{array}{l} H \\ H \\ H \end{array}\right\}N \qquad \left.\begin{array}{l} C_2H_5 \\ H \\ H \end{array}\right\}N \qquad \left.\begin{array}{l} C_2H_5 \\ C_2H_5 \\ H \end{array}\right\}N \qquad \left.\begin{array}{l} C_2H_5 \\ C_2H_5 \\ C_2H_5 \end{array}\right\}N$$

Compounds of the ammonia type

$$\left.\begin{array}{l} H \\ H \end{array}\right\}O \qquad \left.\begin{array}{l} C_2H_5 \\ H \end{array}\right\}O \qquad \left.\begin{array}{l} C_2H_5 \\ C_2H_5 \end{array}\right\}O$$

Compounds of the water type

$$\left.\begin{array}{l} H \\ H \end{array}\right\} \qquad \left.\begin{array}{l} C_2H_5 \\ H \end{array}\right\} \qquad \left.\begin{array}{l} C_2H_5 \\ C_2H_5 \end{array}\right\} \qquad \left.\begin{array}{l} H \\ Cl \end{array}\right\} \qquad \left.\begin{array}{l} C_2H_5 \\ Cl \end{array}\right\}$$

Compounds of the hydrogen type   Compounds of the hydrogen
chloride type

Figure 8.5 Type formulae. These formulae were intended to indicate the chemical similarity of the compounds. They were *not* meant to be structural formulae

Gerhardt, the chief exponent of the new type theory, proposed that all organic compounds could be represented in terms of these four types. It is important to realise that to Gerhardt these formulae were reaction formulae; they indicated that there were certain chemical similarities between compounds of the same type. Gerhardt still maintained that it was impossible to determine the actual arrangement of atoms within a molecule.

One problem with attempting to represent all compounds in terms of four types was that, as the number of known compounds increased, it was necessary to represent them in terms of increasingly complex radicals. This problem was solved to some extent by Williamson, who used *double–type* formulae in which two conventional type formulae were linked by a small radical, which was therefore described as being 'diatomic' or 'dibasic' (Figure 8.6). With the emphasis being placed on smaller polybasic radicals, it was not long before attention was focused on individual atoms, and above all on the atom of the fundamental element in organic chemistry, namely carbon.

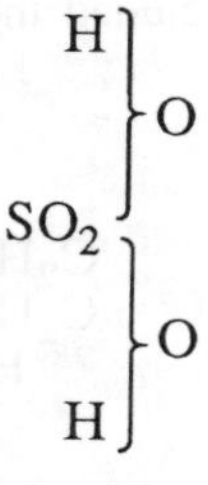

$$
\left.\begin{array}{c} H \\ \\ SO_2 \\ \\ H \end{array}\right\} O
$$

Figure 8.6 Williamson's double–type formula for sulphuric acid. Here the 'dibasic' or 'diatomic' radical SO_2 acts as a kind of molecular cement between the two water types

THE TETRAVALENT CARBON ATOM

The man who introduced the vitally important concept of the tetravalent carbon atom was Friedrich August Kekulé (1829–1896). In 1857, Kekulé proposed that to Gerhardt's four types a fifth should be added, that of marsh gas (methane). In the following year he published a classic paper in which he wrote:

'If only the simplest compounds of carbon are considered (marsh gas, methyl chloride, carbon tetrachloride, chloroform, carbonic acid, carbon disulphide, prussic acid, etc.), it is striking that the amount of carbon that the chemist has known as the least possible, as the *atom*, always combines with four atoms of a monatomic, or two atoms of a diatomic, element; that generally, the sum of the chemical unities of the elements which are bound to one carbon atom is equal to 4. This leads to the view that carbon is *tetratomic* (or tetrabasic).'

Kekulé (Figure 8.7) was born in Darmstadt in Germany, and went to the university of Giessen to study architecture. While there, he attended some of Liebig's lectures and he was so inspired that he switched to chemistry. After leaving Giessen he studied with Dumas in Paris, and from 1853 to 1855 he worked as an assistant at St Bartholomew's Hospital in London. In London he became acquainted with Williamson. In later life he claimed that his theory of the ability of carbon atoms to link to form chains first came to him in a dream that he had on a London omnibus. He also claimed that a dream inspired his theory of the cyclic structure of benzene (Chapter 10). After leaving London he taught first at Heidelberg and then at Ghent. From 1865 he was professor at Bonn, where he remained until his death. He was an outstanding teacher, and three of the first five Nobel Prizes in Chemistry were awarded to his former students: Jacobus van't Hoff (1901), Emil Fischer (1902) and Adolf von Baeyer (1905).

Kekulé's conclusion that the carbon atom was tetratomic is reminiscent of Frankland's theory of the fixed combining powers of nitrogen, phosphorus, arsenic and antimony. The term *valency* was introduced in the 1860s to describe the atomicity or combining power of an element. It is interesting to note that the concept of valency had origins in both contemporary traditions in organic chemistry. Frankland was an adherent of the radical theory, whist Kekulé reached his conclusions as a result of viewing organic compounds in terms of the type theory. The concept of valency forms the basis of modern structural organic chemistry.

Figure 8.7   Friedrich August Kekulé (1829–1896)

Kekulé also proposed that carbon atoms could link together to form chains. When a compound undergoes a reaction in which the number of carbon atoms remains unaltered, he suggested that the carbon skeleton is unaffected and only the atoms joined to it are changed. Thus Kekulé arrived at some of the fundamental ideas of aliphatic chemistry, but the nature of aromatic compounds remained unclear. The best suggestion that he could advance at this stage was that in aromatic compounds the carbon atoms are somehow arranged more densely.

The theory of the tetravalent carbon atom was proposed independently of Kekulé and at almost exactly the same time by the Scotsman Archibald Scott Couper (1831–1892). In some respects Couper's ideas went beyond those of Kekulé, for he considered carbon to display variable valency (e.g. in carbon monoxide and carbon dioxide), and he represented the constitution of carbon compounds by means of graphic formulae (Figure 8.8).

Couper's theory was a more complete break with the past than that of Kekulé. The latter had been an adherent of the type theory and was concerned to explain types in terms of the constant atomicity of an element such as carbon; Couper was intent on replacing the type theory altogether. Unlike Couper, Kekulé did not use graphic formulae in his paper of 1858, and it was not until 1861 that he introduced his system. He represented univalent atoms by single circles, and multivalent atoms by an appropriate number of overlapping circles. These representations appear crude compared to those used by Couper, and even some of Kekulé's German colleagues called them 'sausage formulae' (Figure 8.10). Graphic formulae very similar to those used today were first introduced by the Scotsman Alexander Crum Brown in 1864. Brown and Couper were colleagues for a few months in Edinburgh after the latter's return from France, and Brown's formulae very probably owe something to Couper's influence.

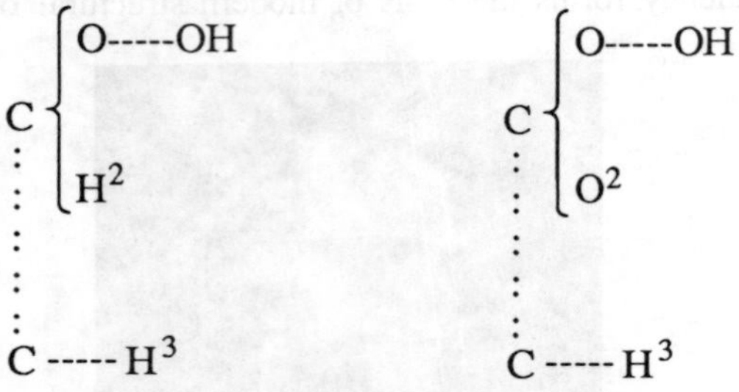

ethyl alcohol   acetic acid

Figure 8.8 Couper's graphic formulae. Couper used the atomic weights C = 12 and O = 8, and in consequence he wrote O—O or O^2 in situations where we would represent a single atom

Couper (Figure 8.9) was born into a prosperous cotton weaving family, and after studying classics and philosophy at the Universities of Glasgow and Edinburgh he travelled extensively in Europe. In 1855 he spent some time in Berlin and started to study chemistry. After a year he moved to Paris to work in Wurtz's laboratory, and it was while he was there that he formed his theory on the constitution of carbon compounds. Couper's appearance on the stage of chemistry was destined to be extremely brief. He had given the paper describing his theory to his superior Wurtz in the expectation that the latter would organise its submission to the French Academy of Sciences. Wurtz was somewhat slow in making the necessary arrangements, and in the meantime Kekulé's paper was published. When Couper took Wurtz to task over the delay, he was judged to be acting insolently and was dismissed from the laboratory. He returned to Scotland and worked for a short while as assistant to Lyon Playfair at Edinburgh, but after a few months he suffered a nervous breakdown and spent the rest of his life in retirement. His important work was totally forgotten until rediscovered after his death by the German chemist Richard Anschütz (1852–1937).

Organic chemistry had made huge advances by 1860 in spite of the continuing uncertainty concerning atomic and molecular weights. Until these problems were resolved there could be no unanimous agreement about either empirical or molecular formulae. However, an event was about to occur which was to end these uncertainties and lead to further rapid progress in both organic and inorganic chemistry.

Figure 8.9 Archibald Scott Couper (1831–1892)

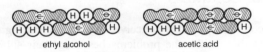

ethyl alcohol   acetic acid

Figure 8.10 Kekulé's graphic formulae

9

The Karlsruhe Congress and its Aftermath

Soon after the publication in 1858 of his classic paper on the tetravalent carbon atom, Kekulé started work on his textbook entitled *Lehrbuch der organischen Chemie*. Organic chemistry had made considerable progress in recent years, but was suffering from the confusion that had been rife ever since the chemical atomic theory had been proposed by Dalton. Atomic weights could not be determined from equivalent or combining weights without a knowledge of the formulae of a few simple compounds. Lack of certainty over atomic weights led to uncertainty in the formulae of all other compounds. Many chemists had turned to equivalent weights instead, and indeed there was confusion over the precise meaning of the terms atom, molecule and equivalent. Gerhardt had suggested a system of atomic weights which was substantially correct, but had not won general acceptance. When Kekulé's book appeared, he quoted nineteen different formulae that had been suggested for acetic acid.

THE KARLSRUHE CONGRESS

As a result of this confusion, Kekulé conceived the idea of calling an international meeting of chemists to consider the problems besetting their subject. In 1859 he discussed this proposal with his friend Carl Welzein (1813–1870), who was professor of chemistry at the Technische Hochschule in Karlsruhe. Welzein undertook to organise an International Chemical Congress in Karlsruhe the following year. Welzein's letter of invitation, sent to all the prominent chemists of Europe, commenced:

'Gentlemen,
Chemistry has reached a state of development when to the undersigned it seems necessary that a meeting of a great number of chemists, active in the science, who are called upon to do research and teach, be held so that a unification of a few important points shall be approached'.

Later in the letter he set out the aims of the proposed meeting:

'More precise definitions of the concepts of atom, molecule, equivalent, atomicity, alkalinity, etc.; discussion of the true equivalents of bodies and their formulas; initiation of a plan for rational nomenclature'.

The Congress convened on 3 September 1860, and the official report shows that 127 of Europe's leading chemists were present. On the third day, one of the subjects discussed was notation. Some younger chemists were following Gerhardt and using atomic weight values of C = 12 and O = 16 in preference to equivalent weights (C = 6, O = 8), which had previously been popular. The Congress was considering whether barred symbols, for example

$$\overline{C} \qquad \overline{O}$$

should be used in formulae to denote when the doubled values were being employed. It was at this point that Stanislao Cannizzaro (1826–1910) made his famous address.

Cannizzaro (Figure 9.1) was born in Palermo in Sicily, and at the age of fifteen he entered the university there to study medicine. He became very interested in chemistry, but because the subject was poorly taught at Palermo he left in 1845 and went to Pisa to work as laboratory assistant to professor Piria, who was the leading Italian chemist of the day. Between 1847 and 1849 he fought in the rebellion staged by Sicily against the King of Naples. After the failure of the revolt, he made his way to Paris and studied with Chevreul. In 1851 he returned to Italy to take up the post of professor at Alessandria, and in 1855 he moved to Genoa. Only a few months before the Karlsruhe Congress he was again involved in political unrest. He went to Palermo in support of Garibaldi, who had taken the city in defiance of the King of Naples, although on this occasion Cannizzaro was not personally involved in the fighting. Soon after the Congress he became professor at Palermo, and in 1871 he accepted the chair in Rome, which he held until his death. In Sicily he worked for the establishment of more schools, and in Rome, where he was made a Senator, he assisted in shaping the Constitution of the recently united Italy.

Cannizzaro spoke in favour of the barred symbols but his contribution was more important for its strong advocacy of Gerhardt's atomic weights. One of the delegates whom Cannizzaro influenced was Mendeleev, who later recalled:

'I vividly remember the impression produced by his speeches, which admitted of no compromise and seemed to advocate truth itself, based on the conceptions of Avogadro, Gerhardt and Regnault which at that time were far from being recognised. And though no understanding could be arrived at, yet the objects of the meeting were attained, for the ideas of Cannizzaro proved, after a few years, to be the only ones which could stand criticism and which represented the atom as 'the smallest portion which enters into a molecule or its compound'. Only such real atomic weights—not conventional ones—could afford a basis for generalisation'.

Figure 9.1 Stanislao Cannizzaro (1826–1910)

As Mendeleev indicated, no agreement was reached at the meeting. However, Cannizzaro had taken the precaution of bringing along some reprints of a paper that he had published two years earlier. As the delegates left the conference, a friend of Cannizzaro's, Angelo Pavesi, distributed the pamphlets. Julius Lothar Meyer (1830–1895) later recorded the impression the paper made on him as he travelled home: 'The scales fell from my eyes, doubts vanished and were replaced by a feeling of the most peaceful assurance'.

Cannizzaro's article was entitled *An Abridgement of a course of Chemical Philosophy given in the Royal University of Genoa*. In the article, Cannizzaro described how in his lecture course he gave an historical account of the origin of the confusion over atomic and molecular weights and explained that a consistent system resulted if Avogadro's hypothesis were accepted. He took the formula of the hydrogen molecule to be H_2 and accorded hydrogen a relative density value of 2 units. Accepting the equal volumes–equal numbers of particles principle, the relative vapour densities of volatile substances on the $H_2 = 2$ scale were then molecular weight values on the $H = 1$ scale.

Cannizzaro then proceeded to show how the molecular weight values thus calculated could be used to justify the assumption that the hydrogen molecule was in fact H_2 (rather than H_4 or H_6, etc.). Using the quantitative analytical data for a number of hydrogen–containing compounds, he calculated the contribution to the molecular weight made by hydrogen. Thus for water, whose molecular weight was 18, he calculated that hydrogen contributed 2 units and oxygen 16. For hydrogen chloride, he calculated that hydrogen contributed only 1 unit to the total value of 36.5. This was the smallest value contributed to any molecular weight by hydrogen, and larger contributions were always whole numbers. Had Avogadro's assumption been wrong, and the true formula of the hydrogen molecule been H_4, then a molecule containing only one atom of hydrogen would have had a contribution of 0.5 to the molecular weight from this element, and in other molecules the contributions due to hydrogen would have been multiples of 0.5.

By similar reasoning Cannizzaro established the atomic weights of other elements. Carbon was shown never to contribute less than 12 units to a molecular weight and oxygen never less than 16. Phosphorus was shown to have an atomic weight of 32, but the molecular weight value of phosphorus in the vapour phase was 124, indicating a P_4 molecule. Mercury vapour was shown to consist of individual atoms, and so the assumption made by Dumas that all elements exist as diatomic molecules in the vapour phase was shown to be false. This had been one of the causes of the confusion in the preceding 30 years.

Although at the time many of the delegates must have thought the Karlsruhe Congress to have been a failure, the inexorable logic of Cannizzaro's paper gradually won converts. Confidence about atomic weight values brought confidence concerning molecular formulae in organic chemistry. Indeed, the confusion reigning before 1860 had contributed greatly to the doubts being expressed about the atomic theory. The new era ushered in by Cannizzaro enabled the atomic theory to survive.

As well as having important consequences for organic chemistry, Cannizzaro's contribution was to have equally far–reaching effects in inorganic chemistry. The growing number of elements had always presented a problem for the atomic theory, but the establishment of reliable atomic weight values enabled useful attempts to be made to arrange the elements in a logical manner.

NEW ELEMENTS

The year of the Karlsruhe Congress also saw the introduction of a technique which resulted in the detection of several new elements. The number of known elements had remained at 58 since 1844, since those as yet unknown were generally present in minerals in quantities too small to be revealed by the analytical techniques then in use. It was with the introduction of the spectroscope by Bunsen and Gustav Robert Kirchhoff (1824–1887) in 1860 that the detection of further elements became possible.

To trace the development of the spectroscope we have to go back to 1758 when Andreas Sigismund Marggraf (1702–1782) had observed that sodium and potassium salts impart different colours to a flame. In the intervening years several observers had analysed the light emitted by metallic salts, finding that it could be resolved by a prism into bright lines separated by dark spaces. In 1814 the Bavarian lens manufacturer Joseph Fraunhofer (1787–1826) observed the solar spectrum through a small telescope, and noticed that the familiar rainbow colours were crossed by a large number of dark lines. He observed similar *Fraunhofer lines* in the spectra of the stars.

Bunsen was investigating the salts in mineral waters, and he decided to use the flame colours for identification purposes. His new laboratory at Heidelberg was equipped with gas, but Bunsen found that the burners available emitted too much light to observe the colours imparted to the flame by the salts. In conjunction with his laboratory assistant he devised the burner named after him, in which air is mixed with gas before combustion occurs. As well as providing a suitably non-luminous flame for Bunsen's immediate needs, the burner became the standard method of laboratory heating, replacing the furnaces and buckets of charcoal then in use.

It was Kirchhoff who suggested to Bunsen that the mixtures present in mineral waters would be better analysed by passing the emitted light through a prism, rather than attempting to remove individual colours by means of filters. Together they designed the first spectroscope (Figure 9.2). Bunsen and Kirchhoff found that the positions of the spectral lines were characteristic of the metal present in the salt and were unaffected by the presence of other salts. They found the method to be of great sensitivity, and announced that it might detect hitherto unknown elements. Kirchhoff also noticed that when intense white light is passed through a sodium flame, its spectrum showed a dark line in the same position as the yellow line in the sodium emission spectrum. The Fraunhofer lines in the spectra of the sun and stars were therefore explained as being due to the absorption of light by elements they contained.

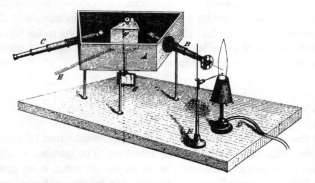

Figure 9.2 Bunsen and Kirchhoff's first spectroscope of 1860

Spectroscopy soon fulfilled its promise, and, using the new technique, Bunsen and Kirchhoff detected and then isolated caesium (1860) and rubidium (1861). Thallium was discovered in 1861 by William Crookes (1832–1919) and indium in 1863 by Ferdinand Reich (1799–1882) and H. T. Richter (1824–1898).

ARRANGING THE ELEMENTS

The first classification of the elements had been made by Lavoisier when he published his table in the *Traité* in 1789 (Chapter 5). Elements were grouped according to the nature of the compound they formed with oxygen. Other arrangements followed, one of which owed something to the Linnaean botanical system of classification. This was the scheme suggested in 1816 by Ampère, who regarded elements as species. Similar species were placed in the same genus, and the genera were grouped into classes. For example, the class *Leucolytes* contained the genus *Calcides,* which comprised the species barium, strontium, calcium and magnesium.

It was only natural that attempts should be made to find relationships between the atomic weight values of chemically similar elements. The first such attempt was made in 1816 by Johann Wolfgang Döbereiner (1780–1849), who noticed that the atomic weight of strontium was close to the mean of the values for calcium and barium. He predicted in his lectures (but not in print) that the as yet unknown atomic weight of bromine would fall midway between those of chlorine and iodine. This prediction was confirmed by Berzelius, and after the latter had published his table of atomic weights in 1826, Döbereiner pointed out that it was possible to find several other groups of three similar elements, or *triads,* displaying the same relationship. Among the examples that he quoted were lithium, sodium and potassium; and sulphur, selenium and tellurium.

More elaborate attempts to find numerical relationships were made independently by Dumas and by the American Josiah Parsons Cooke Jr (1827–1894), who was professor of chemistry at Harvard. In 1854 Cooke found algebraic formulae to link the atomic weights of similar elements. For example, he found that the atomic weights of the elements oxygen, fluorine, chlorine, bromine and iodine conformed to the formula $8 + n9$ where n had the values 0, 1, 3, 8 and 13, respectively. In this pre–Cannizzaro era, Cooke was actually using equivalent weights. Cooke called his groups of similar elements *series,* and likened them to homologous series in organic chemistry. Dumas made the same comparison, noting that for example the atomic (actually equivalent) weights of the similar elements oxygen (8), sulphur (16), selenium (40) and tellurium (64) were all multiples of 8.

Underlying the thinking of Dumas and others was the belief that elements were stable compounds of simpler substances. Dumas suggested that transmutation might be possible after all, and the reason for the failure of the alchemists to produce gold was that they had started with unrelated elements. Faraday also doubted the status accorded to elements. It was the discovery of thallium by Crookes in 1861 that prompted Faraday to remark that a discovery really worth making would be the decomposition of an element (Chapter 7).

With such views being expressed, it was not surprising that there was a revival of interest in Prout's hypothesis, and Dumas proposed that all atomic weights were multiples of 0.25. Just as the careful work of Berzelius had disposed of the notion that atomic weights were multiples of the value for hydrogen, it was the elegant determinations of Jean Servais Stas (1813–1891) that demonstrated that atomic weight values were not exact multiples of 0.25.

The first person to arrange the elements in ascending order of Cannizzaro's atomic weights was the Frenchman A. E. Beguyer de Chancourtois (1820–1886). In 1862 he plotted the atomic weights of the elements in a sloping line on the surface of a cylinder. One complete turn of the cylinder corresponded to 16 units on the atomic weight scale. He found that similar elements fell on or near vertical lines described on the surface of the cylinder, and that helices drawn on the cylinder sometimes brought out other relationships. He called his device the *telluric helix,* but did not illustrate it in his paper. This omission, coupled with the fact that as a geologist he grouped together some elements which to chemists were not particularly alike, resulted in his helix receiving little notice.

At about the same time two British chemists were speculating on the arrangement of the elements. John Alexander Reina Newlands (1837–1898) published several papers between 1863 and 1866. Like Dumas, he initially tried to find numerical relationships between the equivalent weights of the elements, and concluded that the differences between the values for similar elements were multiples of 8. Later he tried ordering all the elements, and found that, if Cannizzaro's atomic weights were used, 'the eighth element, starting from a given one, is a kind of repetition of the first, like the eighth note in an octave of music'. Newlands called his generalisation the *Law of Octaves*. In his table Newlands gave each element a number to indicate its position in the series of increasing atomic weights. Although Newlands left gaps for undiscovered elements in some of his earlier tables, no spaces appeared in the final version of 1866 (Figure 9.3). Some elements were made to share places, and some elements that were not very similar were grouped together. However, he realised that on chemical grounds it was necessary to place tellurium before iodine, in spite of the lower atomic weight of the latter element.

No.		No.		No.		No.		No.		No.		No.		No.	
H	1	F	8	Cl	15	Co & Ni	22	Br	29	Pd	36	I	42	Pt & Ir	50
Li	2	Na	9	K	16	Cu	23	Rb	30	Ag	37	Cs	44	Os	51
G	3	Mg	10	Ca	17	Zn	24	Sr	31	Cd	38	Ba & V	45	Hg	52
Bo	4	Al	11	Cr	19	Y	25	Ce & La	33	U	40	Ta	46	Tl	53
C	5	Si	12	Ti	18	In	26	Zr	32	Sn	39	W	47	Pb	54
N	6	P	13	Mn	20	As	27	Di & Mo	34	Sb	41	Nb	48	Bi	55
O	7	S	14	Fe	21	Se	28	Ro & Ru	35	Te	43	Au	49	Th	56

Figure 9.3 Newlands's table of 1866

Although Newlands's tables clearly represent a foreshadowing of the periodic law, his contemporaries were unimpressed. At a meeting of the Chemical Society at which his ideas were discussed, Newlands was asked if he had ever tried arranging the elements in alphabetical order, since 'any arrangement would present occasional coincidences'. Later, when the periodic law had been accepted, Newlands drew attention to his earlier work and was awarded a medal by the Royal Society.

Meanwhile William Odling (1829–1921) was also arranging the elements in order of ascending atomic weight, and in 1864 he published a table (Figure 9.4), which he introduced with the words:

'With what ease this purely arithmetical seriation may be made to accord with a horizontal arrangement of the elements according to their usually perceived groupings, is shown in the following table, in the first three columns of which the numerical sequence is perfect, while in the other two the irregularities are but very few and trivial'.

This arrangement is fairly similar to the first version of Mendeleev's table published five years later. However, Odling did not enunciate any form of periodic law, but proceeded with another analysis of the relationships between the numerical values of atomic weights of similar elements. He suggested that there was some significance in the fact that the differences between atomic weights of similar elements were usually multiples of 4, or nearly so.

			Ro 104	Pt 197
			Ru 104	Ir 197
			Pd 106·5	Os 199
H 1	„	„	Ag 108	Au 196·5
„	„	Zn 65	Cd 112	Hg 200
L 7	„	„	„	Tl 203
G 9	„	„	„	Pb 207
B 11	Al 27·5	„	U 120	„
C 12	Si 28	„	Sn 118	„
N 14	P 31	As 75	Sb 122	Bi 210
O 16	S 32	Se 79·5	Te 129	„
F 19	Cl 35·5	Br 80	I 127	„
Na 23	K 39	Rb 85	Cs 133	„
Mg 24	Ca 40	Sr 87·5	Ba 137	„
	Ti 50	Zr 89·5	Ta 138	Th 231·5
	„	Ce 92	„	
	Cr 52·5	Mo 96	V 137	
	Mn 55		W 184	
	Fe 56			
	Co 59			
	Ni 59			
	Cu 63·5			

Figure 9.4 Odling's arrangement of the elements of 1864

THE PERIODIC LAW

As the 1860s drew to a close, the idea of periodicity among the elements was definitely 'in the air'. The first clear statement of the periodic law was made in 1869 by Dmitri Ivanovitch Mendeleev (1834–1907). Mendeleev was an outstanding teacher and possessed a very thorough knowledge of the chemical properties of the elements. Soon after taking up his appointment at the University of St Petersburg, he commenced work on his textbook *Principles of Chemistry*. In an attempt to systematise the properties of the elements, he investigated whether chemical behaviour depended in any way on atomic weight. Like some of his contemporaries, he suspected that a correlation might exist, and as he later remarked '... nothing, from mushrooms to a scientific dependence, can be discovered without looking and trying'. He made a series of cards on which he wrote the typical properties and atomic weights of the elements, and by trying different arrangements of the cards he saw that there was a periodic relationship between properties and atomic weights. His first *periodic table* was published in 1869 (Figure 9.5).

			Ti = 50	Zr = 90	? = 180
			V = 51	Nb = 94	Ta = 182
			Cr = 52	Mo = 96	W = 186
			Mn = 55	Rh = 104,4	Pt = 197,4
			Fe = 56	Ru = 104,4	Ir = 198
		Ni = Co = 59		Pd = 106,6	Os = 199
H = 1			Cu = 63,4	Ag = 108	Hg = 200
Be = 9,4	Mg = 24	Zn = 65,2	Cd = 112	
B = 11	Al = 27,4	? = 68	Ur = 116	Au = 197?
C = 12	Si = 28	? = 70	Sn = 118	
N = 14	P = 31	As = 75	Sb = 122	Bi = 210?
O = 16	S = 32	Se = 79,4	Te = 128?	
F = 19	Cl = 35,5	Br = 80	J = 127	
Li = 7 Na = 23	K = 39	Rb = 85,4	Cs = 133	Tl = 204
	Ca = 40	Sr = 87,6	Ba = 137	Pb = 207
	? = 45	Ce = 92		
	?Er = 56	La = 94		
	?Yt = 60	Di = 95		
	?In = 75,6	Th = 118?		

Figure 9.5 Mendeleev's first periodic table of 1869

Figure 9.6 Dmitrii Ivanovitch Mendeleev (1834–1907)

Mendeleev (Figure 9.6) was born in Tobolsk in Siberia, the youngest of fourteen children. His father, who was director of the local high school, started to go blind soon after his son's birth. When his father had resigned his post, his mother reopened a glass factory that her family had previously operated in the neighbourhood. At the age of fifteen, with his father now dead and the glassworks destroyed by fire, Mendeleev's mother took him to Moscow in order to seek entry to the University to study science. He was denied a place, but after travelling to St Petersburg he was admitted to the Central Pedagogic Institute to train as a teacher. Soon afterwards his mother died, but the confidence she had displayed in her youngest child was to prove well placed. Mendeleev graduated in 1855 and after teaching for a year he returned to St Petersburg to work as a research student at the university. His studies on the physical properties of liquids were well received, and in 1859 he was awarded a travelling fellowship, which took him to Paris and Heidelberg. During this time he attended the Karlsruhe Congress. After returning to St Petersburg he was awarded his doctorate, and soon afterwards he was appointed professor of chemistry, first at the Technological Institute, and later at the University. Mendeleev was a man of liberal views, and it was probably because of this that he was never admitted to the Imperial Academy of Sciences. In 1890 there was unrest among the students at the university, and Mendeleev succeeded in ending the disturbances by promising to present the students' grievances to the Minister of Education. Mendeleev received an official rebuke for his conduct and resigned in protest. In 1893 he became Director of the Bureau of Weights and Measures where he remained until his death.

While Odling had apparently failed to notice periodicity and to Newlands it had merely been an interesting curiosity, Mendeleev saw the fundamental significance of the periodic law. As well as presenting his table, Mendeleev made eight statements:

'1. The elements, if arranged according to their atomic weights show a clear periodicity of properties.
2. Elements which are similar as regards their chemical properties have atomic weights which are nearly of the same value (platinum, iridium, osmium) or which increase regularly (potassium, rubidium, caesium).
3. The arrangement of the elements, or of groups of elements, in the order of atomic weights, corresponds with their valencies ....
4. The elements which are most widely distributed in nature have small atomic weights and sharply defined properties. They are therefore typical elements ....
5. The magnitude of the atomic weight determines the character of an element ....
6. The discovery of many new elements may be expected, for example analogues of silicon and aluminium with atomic weights of 65 and 75.
7. Some atomic weights will probably be corrected, for example tellurium cannot have an atomic weight of 128 but somewhere between 123 and 126.
8. The table reveals new analogies between elements ....'

While Mendeleev's prediction of the discovery of new elements analogous to silicon and aluminium was accurate, his suggested correction of the atomic weight of tellurium was not. Tellurium and iodine provide one of the three examples of *inversions* in the periodic table. The reasons for these were not understood until the concepts of atomic number and isotopy were introduced in the twentieth century.

The other chemist associated with the introduction of the periodic table was Lothar Meyer. In 1864, after his conversion to the new system of atomic weights by Cannizzaro, he published an important book, *Die modernen Theorien der Chemie*, which was influential in bringing about acceptance of the new system. The book contained a table of the elements arranged horizontally in terms of increasing atomic weight, and showed that similar elements fell in vertical columns. However, some elements were omitted, and like his contemporaries Lothar Meyer concentrated more on the differences between atomic weights of similar elements than on the principle of periodicity.

Lothar Meyer drew up another table in 1868, but this was not published. In 1870 he announced his own ideas on the subject of periodicity. His paper acknowledged Mendeleev's work, and contained a similar periodic table. However, Lothar Meyer suspected that the value Mendeleev had used for the atomic weight of the recently discovered element indium (75.6) was incorrect. This was double the equivalent weight; Lothar Meyer used 113.4 (three times the equivalent weight), which grouped the element along with boron and aluminium. The more important aspect of Lothar Meyer's paper was that it drew attention to the periodicity of atomic volume. He plotted this quantity (atomic weight/density) as a function of atomic weight (Figure 9.7). This graphical representation of a periodic dependence was readily comprehensible, and provided additional evidence of the validity of the periodic law.

In 1871 Mendeleev published a second paper, which contained a much improved table (Figure 9.8). The elements were arranged horizontally in ascending atomic weight order such that similar elements occurred in *groups* in vertical columns. Each group consisted of two subgroups containing elements similar to each other in the formulae of their oxides. Group VIII remained vacant until iron, cobalt and nickel were added at the end of the fourth row. Copper, silver and gold were listed in group VIII as well as in group I with the alkali metals. Only a few rare–earth elements were known at this time, and in the table they were mixed up with the other heavy elements.

In this paper Mendeleev made some bold and confident assertions. He had previously predicted the existence of two undiscovered elements analogous to aluminium and silicon and lying between zinc and arsenic. He now forecast the existence of a third, similar in properties to boron, and lying between calcium and titanium. He christened the three elements *eka–aluminium*, *eka–silicon* and *eka–boron*, and using the periodic law he made extremely detailed predictions of their physical and chemical properties.

Mendeleev now used the periodic law to correct atomic weight values. Like Lothar Meyer, he reasoned that the atomic weight of indium should be three times its equivalent, and he similarly adjusted the atomic weights of cerium and uranium. Following his statement of 1869 about the atomic weight of tellurium, he now changed its value to 125 to produce atomic weights in sequence. He also altered the values for osmium, iridium, platinum and gold so they fell in that order.

Subsequent atomic weight determinations proved this sequence to be correct, but the value for tellurium remained higher than that for iodine.

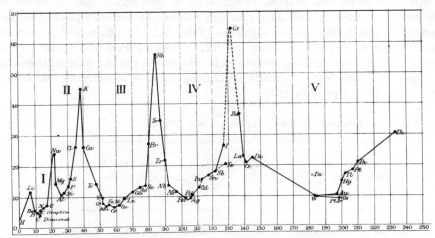

Figure 9.7 Atomic volume plotted against atomic weight (from a publication of 1882; based on Lothar Meyer's curve of 1870)

Reihen	Gruppe I. — R²O	Gruppe II. — RO	Gruppe III. — R²O³	Gruppe IV. RH⁴ RO²	Gruppe V. RH³ R²O⁵	Gruppe VI. RH² RO³	Gruppe VII. RH R²O⁷	Gruppe VIII. — RO⁴
1	H=1							
2	Li=7	Be=9,4	B=11	C=12	N=14	O=16	F=19	
3	Na=23	Mg=24	Al=27,3	Si=28	P=31	S=32	Cl=35,5	
4	K=39	Ca=40	—=44	Ti=48	V=51	Cr=52	Mn=55	Fe=56, Co=59, Ni=59, Cu=63.
5	(Cu=63)	Zn=65	—=68	—=72	As=75	Se=78	Br=80	
6	Rb=85	Sr=87	?Yt=88	Zr=90	Nb=94	Mo=96	—=100	Ru=104, Rh=104, Pd=106, Ag=108.
7	(Ag=108)	Cd=112	In=113	Sn=118	Sb=122	Te=125	J=127	
8	Cs=133	Ba=137	?Di=138	?Ce=140	—	—	—	— — — —
9	(—)	—	—	—	—	—	—	
10	—	—	?Er=178	?La=180	Ta=182	W=184	—	Os=195, Ir=197, Pt=198, Au=199.
11	(Au=199)	Hg=200	Tl=204	Pb=207	Bi=208	—	—	— — — —
12	—	—	—	Th=231	—	U=240	—	

Figure 9.8 Mendeleev's periodic table of 1871

The chemical community was not immediately impressed with the periodic table. It was only with the discovery of the new elements predicted by Mendeleev that its utility began to be appreciated. The first of these elements to be recognised was gallium, found in 1875 by the Frenchman Paul Émile Lecoq de Boisbaudran (1838–1912). He was working with the mineral zinc blende, and once again it was the technique of spectroscopy that revealed the presence of a new element in small quantity. De Boisbaudran ultimately obtained 75 grams of gallium from 4 tonnes of zinc blende. Immediately after the discovery, Mendeleev noted that the properties of gallium that had so far been reported were in agreement with his predictions for eka–aluminium. He wrote: 'If further researches confirm the identity of the properties which I have predicted for eka–aluminium with those of gallium, it will be an instructive example of the utility of the periodic law'. It turned out that there was indeed a remarkable similarity between Mendeleev's postulated eka-aluminium and de Boisbaudran's gallium (Table 9.1). Scandium and germanium were discovered in 1879 and 1886, and their properties were also found to be in accordance with Mendeleev's predictions for eka–boron and eka–silicon. It was the accuracy of Mendeleev's forecasts concerning these new elements which won acceptance for the periodic law.

Table 9.1 Comparison of eka–aluminium and gallium

Properties predicted for eka–aluminium by Mendeleev	*Properties found for de Boisbaudran's gallium*
Atomic weight about 68	Atomic weight 69.9
Metal of specific gravity 5.9; melting point low; non–volatile; unaffected by air; should decompose steam at red heat; should dissolve slowly in acids and alkalis	*Metal* of specific gravity 5.94; melting point 30.15°C; non–volatile at moderate temperature; not changed in air; action of steam unknown; dissolves slowly in acids and alkalis
Oxide: formula Ea_2O_3; specific gravity 5.5; should dissolve in acids to form salts of the type EaX_3. Hydroxide should dissolve in acids and alkalis	*Oxide*: Ga_2O_3; specific gravity unknown; dissolves in acids forming salts of the type GaX_3. The hydroxide dissolves in acids and and alkalis
Salts should have a tendency to form basic salts; the sulphate should form alums; the sulphide should be precipitated by H_2S or $(NH_4)_2S$. The anhydrous chloride should be more volatile than zinc chloride	*Salts* readily hydrolyse and form basic salts; alums are known; the sulphide is precipitated by H_2S and by $(NH_4)_2S$ under special conditions. The anhydrous chloride is more volatile than zinc chloride
The element will probably be discovered by spectroscopic analysis	Gallium was discovered with the aid of the spectroscope

The periodic classification emphasised the importance of atomic rather than equivalent weight, and thereby served to increase confidence in the chemical atomic theory. On the other hand, the fact that there was some kind of relationship between chemical properties and atomic weight, even if not of a precise mathematical nature, helped once again to arouse interest in Prout's hypothesis. It was in order to provide a final test for Prout's hypothesis that the Englishman Lord Rayleigh (1842–1919) embarked in 1882 on a series of very accurate determinations of the atomic weights of gaseous elements by careful measurement of their densities. The unexpected outcome of his research was the addition of a completely new group to the periodic table.

THE DISCOVERY OF THE RARE GASES

In 1892 Rayleigh reported that he had found that oxygen was 15.882 times more dense than hydrogen, and that this value was constant irrespective of how the oxygen had been prepared. However, when he measured the density of nitrogen, he found that samples of the gas derived from the atmosphere had a slightly higher density than 'chemical' nitrogen prepared from ammonia. Rayleigh initially thought that atmospheric nitrogen contained a proportion of N_3 molecules analogous to the ozone allotrope of oxygen, O_3. Ozone is prepared by passing a silent electrical discharge through oxygen, but when Rayleigh performed a similar experiment with nitrogen he found that its density remained unchanged. Another suggestion of Rayleigh's was that chemical nitrogen might have become slightly dissociated into free atoms during the course of its preparation.

At this stage William Ramsay (1852–1916) started to work on atmospheric nitrogen to see if it contained a previously undetected component. He passed atmospheric nitrogen repeatedly through heated magnesium, thus forming magnesium nitride (Figure 9.9). When the residual gas was examined spectroscopically it still exhibited the bands characteristic of nitrogen, but also showed some new lines that did not belong to any known element. Further chemical treatment removed the last traces of nitrogen. Meanwhile Rayleigh had repeated Cavendish's experiment of sparking atmospheric nitrogen with oxygen over alkali and had also obtained a residue of an unreactive gas displaying the novel emission spectrum.

The new gas was found to be almost twenty times more dense than hydrogen. Ramsay found that the ratio of its specific heat capacities was 1.66, indicating that the gas was monatomic, and it therefore followed that its atomic weight was nearly 40. All attempts to induce the new gas to combine chemically met with failure.

Rayleigh and Ramsay presented a preliminary report of their findings in 1894, and gave a fuller account the following year. They christened the new gas *argon* from the Greek word meaning *idle*. Although the evidence pointed strongly to argon being a new element, there remained the considerable problem of how to accommodate it in the periodic table. An atomic weight of 40 pointed to a position between potassium and calcium, where no vacancy existed, and even if there had been a space it would not have been filled by an unreactive monatomic gas. These difficulties prompted several chemists, including Mendeleev, to argue that, in spite of the specific heat evidence, argon was an triatomic allotrope of nitrogen as Rayleigh had originally suggested.

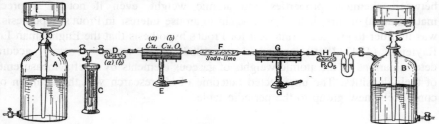

Figure 9.9 Ramsay's apparatus for the isolation of argon. A sample of atmospheric nitrogen was passed backwards and forwards through red–hot magnesium in tube G. The gas was also passed through red–hot copper oxide to oxidise any suspended carbonaceous matter such as dust to carbon dioxide. This was then absorbed by the soda–lime

Soon afterwards Ramsay discovered a second new gas by heating the mineral cleveite. The spectrum of this gas showed a line coincident with a Fraunhofer line that been observed in the solar spectrum in 1868 by the astronomer Joseph Norman Lockyer (1866–1920). Since the line was not shown by any element known on earth, Frankland had named it *helium* after the Greek *helios* meaning sun. Ramsay found that his terrestrial helium was as unreactive as argon.

The problems of classification posed by the two gases were not cleared up until three more were discovered. Ramsay predicted that if helium and argon were genuine elements with atomic weights of approximately 4 and 40, another new gas with an intermediate atomic weight should exist. In 1898 he and Morris William Travers (1872–1961), by careful fractionation of liquid air, discovered krypton, neon and xenon. These three elements were shown to have atomic weights that fell between the halogens and the alkali metals. It was therefore easier to accept that the atomic weights of potassium (39.1) and argon (39.9) were inverted, as was already suspected for tellurium and iodine, and for cobalt and nickel. In 1900 Ramsay and Travers suggested that the new gases, with the possible exception of helium, be placed in a new group in the periodic table. Mendeleev gave his approval to this scheme in 1905 when he placed all five gases in 'group 0', preceding the alkali metals. He had abandoned the belief that argon was N_3, but gave the element an atomic weight of 38. He never accepted atomic weight inversions, believing that they would ultimately be eliminated by more accurate determinations.

FURTHER PROBLEMS OF CLASSIFICATION

The classification of the rare earth elements continued to be a problem into the twentieth century. It was not until Moseley had published his work on X–ray spectra and atomic numbers (Chapter 11) that it was possible to predict how many of these very similar elements should exist. It is now usual to place the rare earths, or *lanthanides*, in an inner transition series in the table, and this is also done with their congeners, the *actinides*.

When the phenomenon of radioactivity was being investigated early in the twentieth century, it became clear that samples of elements could be obtained that were chemically identical but which nevertheless possessed different atomic weight values. Frederick Soddy (1877–1956) named these species *isotopes*, from the Greek meaning *same place*, as they clearly occupied the same position in the periodic table (Chapter 11).

Since the 1880s the periodic table has become the cornerstone of inorganic chemistry. Many versions have been produced, and new ones are still making their appearance. Tables based on the one Mendeleev published in 1871 are said to be of the *short form*. Nowadays the *long form* is more popular; in this the main and subgroup elements are separated (Figure 9.10).

Only eleven years after the Karlsruhe Congress, the new atomic weights were widely accepted, organic chemists were beginning to agree on molecular formulae, and Mendeleev's improved periodic table had been introduced. Cannizzaro's advocacy of atomic weights based on the acceptance of Avogadro's hypothesis can truly be said to be a turning point in nineteenth century chemistry. Chemists were now able to start thinking about molecular structure, and in consequence there were soon to be further important developments in organic chemistry.

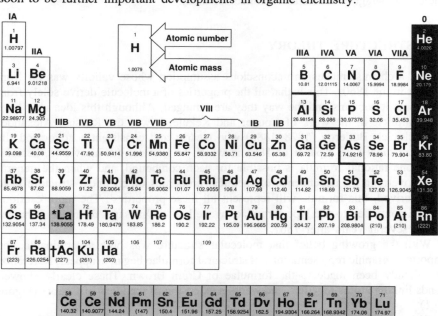

Figure 9.10 A modern periodic table (long form)

10

Organic Chemistry since 1860

In 1858, Couper and Kekulé advanced the theory of the tetravalent carbon atom and the idea that carbon atoms could link to themselves to form chains. Gerhardt, the great exponent of the type theory who had died in 1856, had always maintained that the structure of a molecule could never be known. However, in consequence of the work of Couper and Kekulé, this view was now being questioned, and as a result of the growing body of knowledge concerning organic compounds, chemists could now suggest structural formulae for them.

THE STRUCTURE THEORY

Today, a chemist makes an unconscious assumption whose validity was far from evident in the 1860s. This is that all the properties of a molecule derive solely from the atoms it contains and the way they are arranged. Although this idea may have been implicit in the work of Couper and Kekulé, it was the Russian Alexander Mikhailovich Butlerov (1828–1886) who clearly and consistently advocated the *structure theory*. In a paper of 1861 in which he first set out his views, he said: 'Only one rational formula is possible for each compound, and when the general laws governing the dependence of chemical properties on chemical structure have been derived, this formula will represent all these properties'. The more cautious Kekulé continued to use type formulae, as well as his 'sausage' formulae (Chapter 8), for several years after 1861.

With the growing belief that molecular structure was both ascertainable and important, graphic representations of structural formulae began to appear. Reference has already been made to the formulae of Crum Brown. These clearly showed bonds linking atoms together, and were an extension of the ideas of Couper (Figure 10.1).

Objections were raised to Crum Brown's formulae on the grounds that they appeared to suggest that all the atoms in the molecule were coplanar, in spite of his insistence that he 'did not mean to indicate the physical, but merely the chemical position of the atoms'. The formulae became established fairly quickly in England, partly because Frankland adopted them in his book *Lecture Notes for Chemical Students*, published in 1866. In 1865 Hofmann made some models, based on Crum Brown's formulae, from croquet balls and rods, which he used to illustrate some lectures he gave at the Royal Institution that year (Figure 10.2).

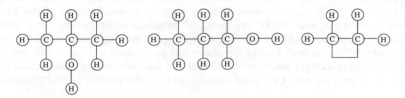

Figure 10.1 Crum Brown's structural formulae

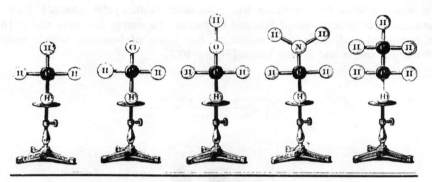

Figure 10.2 Hofmann's croquet ball models

In 1869 Vladimir Vasil'evich Markovnikoff (1838–1904), a pupil of Butlerov, proposed the rule subsequently named after him concerning the addition of hydrogen halides to unsaturated hydrocarbons. In his words: 'If an unsymmetrically constituted hydrocarbon unites with a hydrohalogen acid, the halogen adds to the less hydrogenated carbon atom ...'. This rule provided the first prediction of the course of a reaction in terms of the structure of one of the reactants.

THE BENZENE RING

Unsaturated aliphatic hydrocarbons presented few problems in terms of structure and bonding; Crum Brown had shown a double bond in ethene in 1864. However, the structure of benzene presented a serious difficulty, and it was here that Kekulé was to make another vital contribution to the development of organic chemistry.

Later in life Kekulé claimed that both the concept of the ring structure of benzene and that of the self–linking ability of carbon atoms had occurred to him in the course of dreams. Visions of carbon atoms joining to one another had appeared in a reverie on the top of a London omnibus, while the benzene structure was the consequence of a doze at home:

'I was sitting, writing at my text–book; but the work did not progress; my thoughts were elsewhere. I turned my chair to the fire and dozed. Again the atoms were gambolling before my eyes. This time the smaller groups kept modestly in the background. My mental eye, rendered more acute by repeated visions of this kind, could now distinguish larger structures, of manifold conformation: long rows, sometimes more closely fitting together; all twining and twisting in snake–like motion. But look! What was that? One of the snakes had seized hold of its own tail, and the form whirled mockingly before my eyes. As if by a flash of lightning I awoke; and this time also I spent the rest of the night working out the consequences of the hypothesis'.

Kekulé's first paper on the cyclic structure of benzene appeared in 1865. The structure was represented by means of one of his 'sausage' formulae. Each long thin carbon atom was shown to overlap with its neighbour on one side by two affinity units, and with that on the other side by only one. The spare affinity units at either end were satisfied by combining with each other in the cyclic structure. Later in the same year he represented benzene by means of a simple hexagon, and in 1866 he published a diagram of a model he had made of benzene, which showed alternating single and double bonds (Figure 10.3).

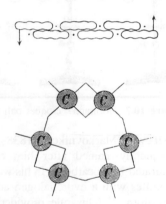

Figure 10.3 Kekulé's representations of the benzene structure

Objections were soon raised that benzene did not show the reactivity which would have been predicted for a compound showing three double bonds (i.e. cyclohexatriene). Furthermore, such a view of benzene leads to the prediction that two 1,2–disubstituted benzenes should exist (Figure 10.4). In practice, it was found that three disubstituted benzenes could exist, but when methods were devised for assigning structures to them (see below), only one was shown to be a 1,2–isomer.

Although these criticisms might seem to us to be very serious, to the chemists of the day the hexagonal structure for benzene solved so many problems that it was widely accepted. The reason for the equivalence of the carbon–carbon bonds and for the lack of reactivity remained matters for speculation, but the fundamental concept of the hexagonal ring encountered little opposition. The only serious alternative was the prism formula of Albert Ladenburg (1842–1911), but, while this could explain the number of di- and trisubstituted isomers, it could not account for the formation from benzene of dihydro and tetrahydro derivatives that clearly contained double bonds (Figure 10.5).

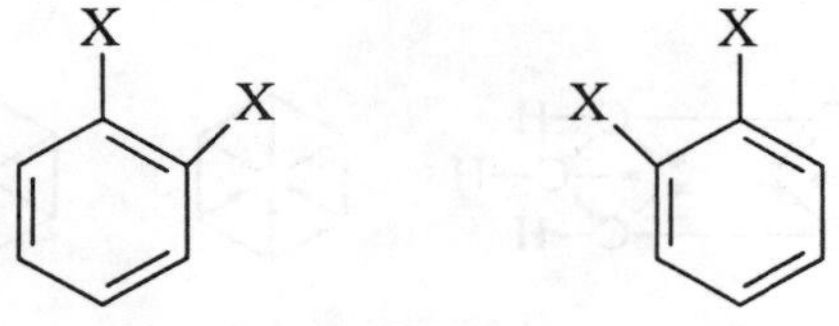

Figure 10.4 The two 1,2–disubstituted benzenes that were to be expected if the benzene ring contained three localised double bonds

Figure 10.5 Ladenburg's prism formula for benzene

The problem of bonding in benzene has continued to exercise chemists until the present time. Kekulé put forward an ingenious suggestion in 1872 that the atoms in the molecule were oscillating rapidly and colliding with their neighbours. The number of collisions made by an atom in unit time was equal to its valency. Thus in Figure 10.6, C^1 in one oscillation might collide with C^2, C^6, H, C^2, but in the course of the next oscillation the sequence might be reversed to give C^6, C^2, H, C^6. This would mean that in the first cycle the double bond would be located betweeC^1 and C^2, and in the second between C^1 and C^6. The concept of valency being related to atomic collisions was not generally accepted, but the idea of oscillating double bonds was frequently employed, and can be regarded as the forerunner of the modern concept of resonance.

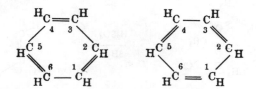

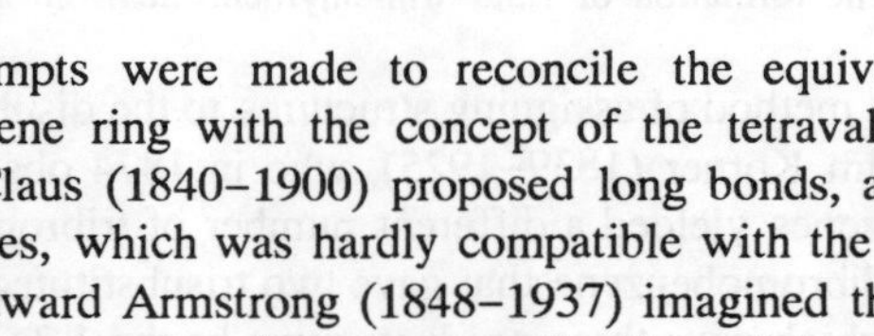

Figure 10.6 Kekulé's suggestion of 1872 concerning the bonding in benzene

Several other attempts were made to reconcile the equivalence of all six positions in the benzene ring with the concept of the tetravalent carbon atom. Adolf Carl Ludwig Claus (1840–1900) proposed long bonds, and Lothar Meyer envisaged free affinities, which was hardly compatible with the lack of reactivity of benzene. Henry Edward Armstrong (1848–1937) imagined that a carbon atom influenced each of the other five through centrally directed bonds, and Adolf von Baeyer (1835–1917) put forward a somewhat similar idea the following year. In 1867, James Dewar (1842–1923) exhibited some models designed to illustrate the possible structures of some unsaturated hydrocarbons. He made seven models of benzene, some of them decidedly bizarre, but they contained the structure which has since borne his name (Figure 10.7).

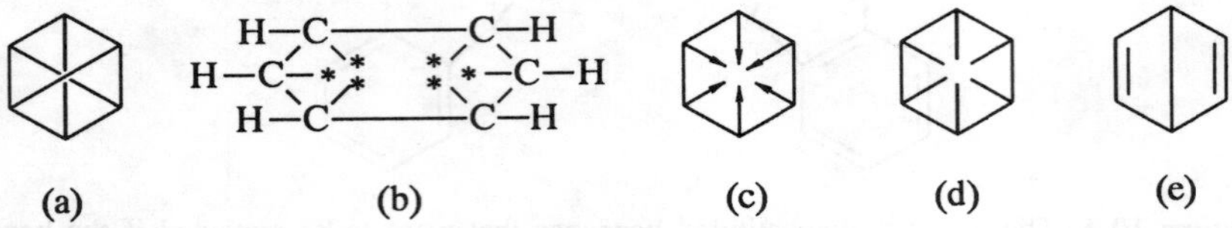

(a)   (b)   (c)   (d)   (e)

Figure 10.7 Various proposed structures for benzene: (a) Claus, (b) Lothar Meyer, (c) Armstrong, (d) Baeyer and (e) Dewar

STRUCTURAL FORMULAE OF BENZENE DERIVATIVES

The hexagonal ring structure of benzene provided an explanation of the occurrence of three disubstituted benzenes, but there remained the considerable problem of assigning a structure to each of the three isomers. In 1867 Carl Graebe (1841–1927) argued that, of the benzenedicarboxylic acids, the one which readily formed an anhydride (phthalic acid) must have the two carboxyl groups on adjacent carbon atoms.

In 1866 Fittig synthesised a trimethylbenzene from acetone, and in 1874 Baeyer argued that the product was 1,3,5–trimethylbenzene in view of its method of preparation (Figure 10.8). Since it was possible to demethylate this trimethyl-benzene to give a xylene, it was deduced that this xylene was 1,3–dimethyl-benzene, and the dicarboxylic acid obtained when it was oxidised was benzene–1,3–dicarboxylic acid (isophthalic acid).

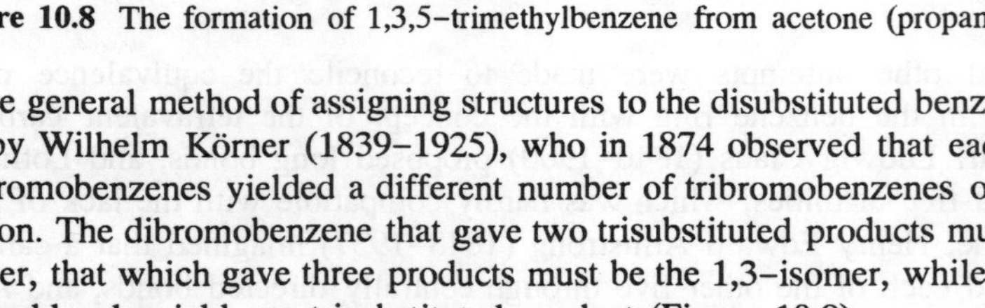

Figure 10.8 The formation of 1,3,5–trimethylbenzene from acetone (propanone)

A more general method of assigning structures to the disubstituted benzenes was devised by Wilhelm Körner (1839–1925), who in 1874 observed that each of the three dibromobenzenes yielded a different number of tribromobenzenes on further substitution. The dibromobenzene that gave two trisubstituted products must be the 1,2–isomer, that which gave three products must be the 1,3–isomer, while the 1,4–isomer could only yield one trisubstituted product (Figure 10.9).

By the mid–1870s, as a result of Cannizzaro's work, chemists were gaining confidence that they were assigning correct molecular formulae to compounds. In consequence of arguments based largely on methods of synthesis, they were increasingly able to write structural formulae. Only a few, most notably Kolbe, ridiculed the confidence with which chemists were writing structures for particles which were unimaginably small.

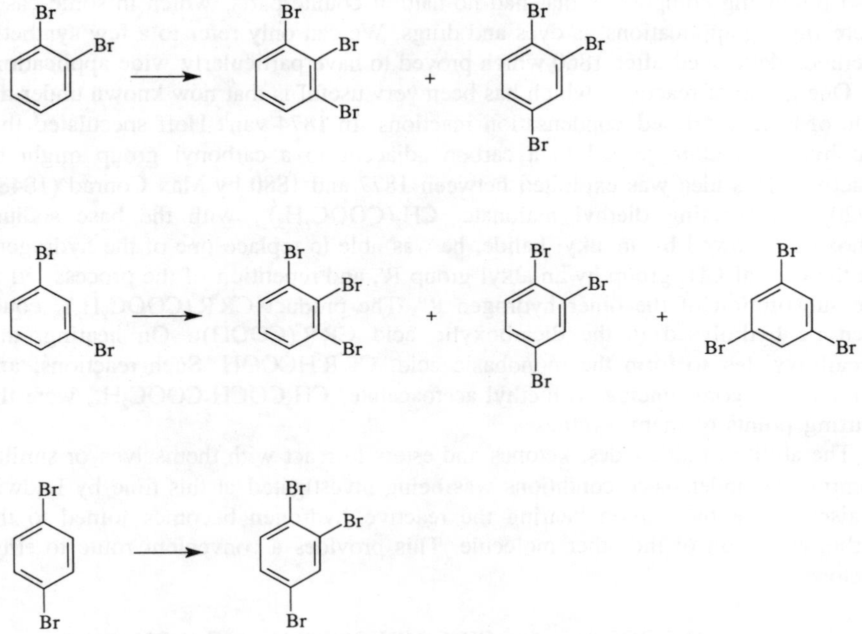

Figure 10.9 Körner's deduction of the structures of the dibromobenzenes on the basis of the number of trisubstituted products that each yielded.

ORGANIC SYNTHESIS 1860–1900

We have seen how molecular structures were being assigned to organic molecules on the basis of their method of synthesis or by considering the products which they yielded on further reaction. The year 1860 marked a turning point in synthetic organic chemistry. In that year, the Frenchman Marcelin Berthelot (1827–1907) published his book *Chimie organique fondée sur la synthèse*. One of Berthelot's aims was to demonstrate that organic compounds of many types could be made from inorganic precursors. Berthelot was the first to use the word 'synthesis', and to him it meant the demonstration of how a given organic compound could be made from its elements. He was therefore extending the work of Kolbe when the latter had prepared acetic acid (Chapter 8). Berthelot's book was the first comprehensive treatise on synthetic organic chemistry, and brought together the general methods discovered thus far for the preparation of many types of compound. As such, it provided a stimulus for further synthetic work at a time when the structure theory was gaining general acceptance.

The huge increase in synthetic organic chemistry after 1860 was not, of course, solely directed either at preparing naturally occurring compounds from inorganic starting materials or at providing evidence on structural questions. Chemists were also producing compounds that had no natural counterparts, which in some cases were finding applications as dyes and drugs. We can only refer to a few synthetic methods developed after 1860 which proved to have particularly wide application.

One group of reactions which has been very useful is that now known under the title of base–catalysed condensation reactions. In 1874 van't Hoff speculated that the hydrogen atom joined to a carbon adjacent to a carbonyl group might be reactive. This idea was exploited between 1877 and 1880 by Max Conrad (1848–1920). By reacting diethyl malonate, $CH_2(COOC_2H_5)_2$, with the base sodium ethoxide followed by an alkyl halide, he was able to replace one of the hydrogens on the central CH_2 group by an alkyl group R', and repetition of the process led to the substitution of the other hydrogen R''. The product $CR'R''(COOC_2H_5)_2$ could then be hydrolysed to the dicarboxylic acid $CR'R''(COOH)_2$. On heating, this decarboxylates to form the monobasic acid, $CR'R''HCOOH$. Such reactions, and similar ones commencing with ethyl acetoacetate, $CH_3COCH_2COOC_2H_5$, were the starting points of many syntheses.

The ability of aldehydes, ketones and esters to react with themselves or similar compounds under basic conditions was being investigated at this time by Ludwig Claisen. Here the carbon bearing the reactive hydrogen becomes joined to the carbonyl carbon of the other molecule. This provides a convenient route to ethyl acetoacetate:

$$2CH_3COOC_2H_5 \rightarrow CH_3COCH_2COOC_2H_5 + C_2H_5OH$$

An important reaction which enabled many new aromatic compounds to be prepared was that discovered in 1877 by Charles Friedel (1832–1899) and James Mason Crafts (1839–1917). They found that an alkyl or acyl group could be substituted into a benzene ring by the reaction of the aromatic hydrocarbon with an alkyl or acyl halide in the presence of aluminium chloride catalyst. The reaction proved to be of wide application and is particularly useful in the preparation of aromatic ketones and in the formation of a new ring by an intramolecular process (Figure 10.11).

There had been a number of attempts to utilise organometallic compounds as synthetic intermediates following the discovery of diethylzinc by Frankland in 1849 (Chapter 8). In 1875, the Russian Alexander Mikhailovich Saytzeff (1841–1910) discovered a method of synthesising primary and secondary alcohols from esters, ketones and aldehydes by the action of zinc and an alkyl iodide. In 1899, Philippe Antoine Barbier (1848–1922), professor of chemistry at Lyons, reported that an attempt to use the Saytzeff reaction to convert methylheptenone to dimethylheptenol by means of zinc and iodomethane was unsuccessful. However, some product was obtained if the zinc was replaced by magnesium. Barbier gave his research student Victor Grignard (1871–1935) (Figure 10.10) the problem of investigating the reaction further, and very soon Grignard had demonstrated the enormous possibilities of alkyl– and phenylmagnesium halides (later known as *Grignard reagents*) in synthesis. In his original paper of 1900, Grignard described how he had prepared methylmagnesium iodide by the reaction of magnesium with methyl iodide dissolved in anhydrous ether. He then added an aldehyde or ketone to the resulting ethereal solution, and obtained a secondary or tertiary alcohol in good yield by hydrolysing the resulting intermediate:

$$CH_3I + Mg \rightarrow CH_3MgI$$
$$CH_3MgI + RCHO \rightarrow RCH(OMgI)CH_3$$
$$RCH(OMgI)CH_3 + H_2O \rightarrow RCH(OH)CH_3 + MgIOH$$

Figure 10.10 Victor Grignard (1871–1935)

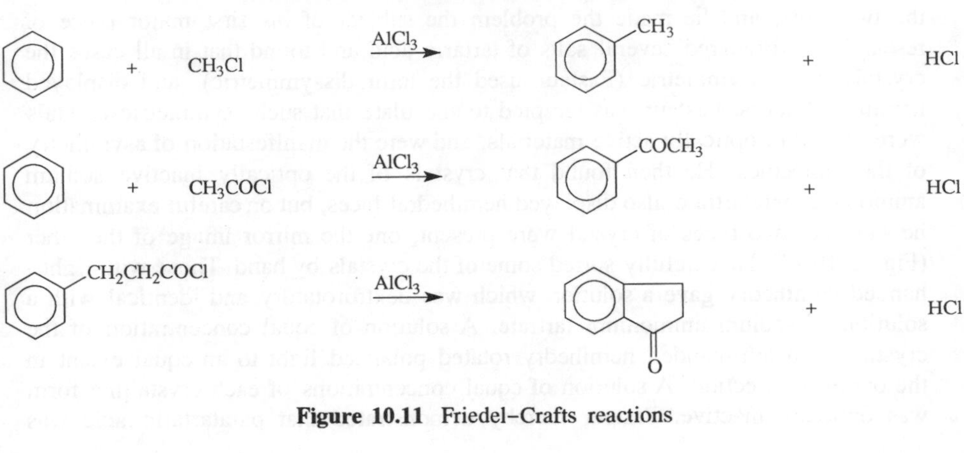

Figure 10.11 Friedel–Crafts reactions

It was soon found that Grignard reagents react with other classes of compound. With carbon dioxide a carboxylic acid is produced, and a hydrocarbon is formed when a Grignard reagent reacts directly with water. Grignard's work had an immediate effect on organic synthesis, and over 500 papers on organomagnesium compounds appeared within eight years of his original publication. In 1912 Grignard was awarded the Nobel Prize in Chemistry. He created a certain amount of controversy by protesting that Barbier should also have been honoured for his part in the discovery.

OPTICAL ACTIVITY

In 1813 Jean Baptiste Biot (1774–1862) had discovered that quartz crystals could rotate the plane of plane–polarised light, and two years later he found that the same property was displayed by various natural organic products such as the liquids oil of turpentine and oil of lemon, and also by a solution of camphor in ethanol. This property, known as *optical activity*, was found to disappear in the case of quartz when the crystals were fused. Augustin Jean Fresnel (1788–1827) suggested that the optical activity of the quartz was due to a helical arrangement of the molecules in the crystal, but it was evident that when the property was displayed by substances in the liquid or solution phase the explanation must reside in the structure of the molecule itself.

The scientist whose name is most closely associated with early work on optical activity is Louis Pasteur (1822–1895). In 1844, while still a student in Paris, Pasteur read a recently published paper by Mitscherlich on the salts of the isomeric tartaric and paratartaric acids. The only known difference in the properties of the two acids and their salts was that the tartrates were optically active while the paratartrates were not. Mitscherlich reported that crystals of the sodium ammonium salts of tartaric and paratartaric acids were isomorphous and he therefore concluded that 'the nature and the number of atoms, their arrangement and their distances, are the same in the two substances compared'.

Pasteur was convinced that there must be some molecular difference between the two salts, and he made the problem the subject of his first major piece of research. He prepared several salts of tartaric acid and found that in all cases the crystals were asymmetric (Pasteur used the term dissymmetric), and displayed hemihedral faces. Pasteur was tempted to speculate that such asymmetric crystals were typical of optically active materials, and were the manifestation of asymmetry of the molecules. He then found that crystals of the optically inactive sodium ammonium paratartrate also displayed hemihedral faces, but on careful examination he saw that two types of crystal were present, one the mirror image of the other (Figure 10.13). He carefully sorted some of the crystals by hand. Those with right–handed hemihedry gave a solution which was dextrorotatory and identical with a solution of sodium ammonium tartrate. A solution of equal concentration of the crystals with left–handed hemihedry rotated polarised light to an equal extent in the opposite direction. A solution of equal concentrations of each crystalline form was optically inactive. Pasteur thereby demonstrated that paratartaric acid was composed of equal quantities of tartaric acid (dextrorotatory) and the new laevorotatory acid. He proposed that since the two isomers formed crystals which were mirror images of each other, the two molecules were also related in the same manner.

Pasteur (Figure 10.12) was born at Dole in France. In 1848, after studying in Paris, he became professor of physics at the Lycée in Dijon. In 1852 he was appointed professor of chemistry at Strasbourg, and in 1852 he became Director of the École normale in Paris. His study of tartaric acid and the tartrates was the only purely chemical piece of research he carried out. Most of his subsequent work was concerned with microorganisms; indeed, he was the founder of the science of microbiology. He announced his germ theory of fermentation in 1857, and it aroused considerable interest because at the time fermentation was thought to be a purely chemical process. He also attacked the widely held theory of spontaneous generation of microorganisms, showing that the phenomena attributed to this cause could always be explained by contamination. He introduced improvements into wine, beer and vinegar production, and invented a method for preventing food spoilage (pasteurisation). His last studies were on diseases of animals and man, and he devised methods of attenuating the virulence of an infective organism so that vaccines could be prepared. By the end of his life Pasteur had acquired an enormous reputation, and the French government established the famous Institut Pasteur in 1888. After his death, Pasteur was interred in the Institut.

Figure 10.12 Louis Pasteur (1822–1895)

Further work by Pasteur showed that it was also possible to separate or *resolve* the two isomers present in paratartaric acid by forming a salt with a naturally occurring optically active base such as *l*–cinchonidine. The two salts had markedly different solubilities and could be separated by fractional crystallisation. A third method, also discovered by Pasteur, was to use a microorganism which consumed one of the enantiomers but not the other. Thus he found that the mould *Penicillium glaucum* grown in a solution of ammonium paratartrate containing a little phosphate gradually removed the *d*–tartrate, leaving the *l*–tartrate unaffected.

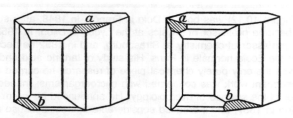

Figure 10.13 The two mirror image crystals formed by sodium ammonium paratartrate. The hemihedral faces are labelled *a* and *b*.

Pasteur converted *d*–tartaric acid into paratartaric acid (and hence to the *l*–form) by heating the *l*–cinchonidine salt to 170°C and extracting the resulting mass with water. He also found that a fourth form of the acid, mesotartaric acid, was formed in this process. This was optically inactive but could not be resolved into enantiomers.

Pasteur contended that the ability to produce an asymmetric molecule was a property of living organisms, and that an optically active material could never be obtained from an inactive precursor in the laboratory. He thought that all compounds displaying optical activity would have an inactive *meso* isomer, and that laboratory syntheses would result in such forms. He was forced to abandon this view in 1860 when W. H. Perkin (1838–1907) and B. F. Duppa made paratartaric acid from succinic acid.

Pasteur was fortunate that he conducted his researches on the salts of tartaric acid. Optical enantiomers do not generally produce asymmetric crystals. Furthermore, Pasteur must have obtained his crystals of sodium ammonium paratartrate from a cool solution. Below 27°C the two enantiomers form separate crystals, but at higher temperatures symmetrical crystals are obtained containing equal quantities of the two forms. Commenting on the good fortune that accompanied several of his researches, Pasteur said 'chance favours only the prepared mind'.

In 1860 Pasteur had speculated on the cause of molecular asymmetry:

'Are the atoms of the dextro–acid grouped on the spiral of a right handed helix, or are they situated at the apices of a regular tetrahedron, or are they disposed according to some other asymmetric arrangement? We do not know. But of this there is no doubt, that the atoms possess an asymmetric arrangement like that of an object and its mirror image'.

STEREOCHEMISTRY

The question posed by Pasteur was answered in 1874. In that year, Jacobus Henricus van't Hoff (1852–1911) and Joseph Achille Le Bel (1847–1930) published papers within two months of each other in which they proposed that optically active molecules contained a carbon atom joined by tetrahedrally directed bonds to four different atoms or groups. They further showed that the tetrahedral model explained why only one isomer existed of molecules of the type $CH_2R'R''$, whereas two would be predicted if all four bonds were directed in the same plane.

Van't Hoff extended his discussion to molecules containing a carbon–carbon double bond. In his diagrams, he represented such molecules by two tetrahedra with an edge in common (Figure 10.14). He showed that the isomerism of maleic and fumaric acids could be explained by both molecules possessing the formula $CHCOOH\!\!=\!\!CHCOOH$, but in one case the carboxyl groups would both lie on the same side of the double bond, and on opposite sides in the other. Referring to previous work on these compounds, van't Hoff said 'all previous explanations of the isomerism between these have made shipwreck'. Van't Hoff concluded by observing that compounds containing the carbon–carbon triple bond could be represented by two tetrahedra with a face in common.

Van't Hoff's paper was published in Dutch, and soon appeared as a French translation entitled *La Chimie dans l'Espace*. Kolbe, always suspicious of molecular structures, was quick to attack the new theory:

'A certain Dr J. H. van't Hoff, of the veterinary school of Utrecht, finds, so it seems, no taste for exact chemical investigation. He has thought it more convenient to mount Pegasus (obviously loaned by the veterinary school) and to proclaim in his *La Chimie dans l'Espace* how, during his bold flight to the top of the chemical Parnassus, the atoms appear to have grouped themselves in universal space'.

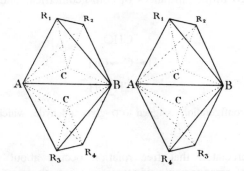

Figure 10.14 Van't Hoff's representation of double–bonded compounds

Other explanations were advanced for molecular asymmetry, some of which referred to the shapes of the atoms themselves. Nevertheless, the concept of the tetrahedrally directed valencies of carbon slowly gained ground during the last quarter of the nineteenth century, and it was universally accepted by the time direct experimental proof was furnished by X-ray diffraction studies on diamond in 1913 by W. H. Bragg (1862–1942) and W. L. Bragg (1890–1971).

After van't Hoff published his theories on the tetrahedral carbon atom and on optical and geometric isomerism in 1874, much work was done to confirm his ideas by Johannes Adolf Wislicenus (1835–1902). He showed how the chemistry of maleic and fumaric acids was best explained if they were geometric isomers. The isomerism of Pasteur's tartaric acid was explained by proposing that it possessed two asymmetric carbon atoms, and that the *meso* form was inactive by internal compensation. Paratartaric acid was also known as *racemic acid* since it sometimes crystallises out from wine (the Latin *racemus* means a bunch of grapes). Inactive mixtures of optical antipodes are now known as *racemic mixtures*.

A consequence of the theory of the tetrahedral carbon atom was that the bond angle at a single–bonded carbon atom should be 109°28'. In 1885 Adolf von Baeyer (1835–1917) introduced his *strain theory*, which stated that five– and six–membered rings were stable compared to rings of other sizes because the distortions from the tetrahedral angle were smallest in these cases. Baeyer was making the assumption that the carbon atoms in cyclic compounds were all coplanar.

As a result of Emil Fischer's work on sugars (see below), it became the custom to relate the absolute configuration of optical isomers to D–glyceraldehyde (Figure 10.15). The configuration of D–glyceraldehyde had been assigned on an arbitrary basis, but in 1951 the X–ray studies of J. M. Bijvoet showed that this configuration was correct. In the twentieth century other examples of optical isomerism have been reported where the molecular asymmetry is not simply caused by asymmetric carbon atoms. Some examples are depicted in Figure 10.16.

The instances of geometric isomerism also multiplied with the realisation that *cis* and *trans* isomerism was to be found in double–bonded nitrogen compounds, and also in saturated cyclic compounds. Werner's work on inorganic coordination compounds provided further instances of both geometrical and optical isomerism (Chapter 12).

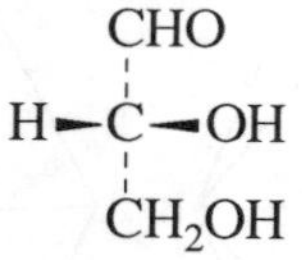

$$
\begin{array}{c}
CHO \\
| \\
H \!-\! C \!-\! OH \\
| \\
CH_2OH
\end{array}
$$

Figure 10.15 The configuration assigned to D–glyceraldehyde, which Bijvoet showed to be correct in 1951

Van't Hoff appreciated that free rotation occurs about single bonds. The consequent different arrangements of atoms in space were termed the *conformations* of the compound by Walter Norman Haworth (1883–1950) in 1929. In 1936 K. S. Pitzer pointed out that there would be an energy barrier to rotation about the carbon–carbon bond in ethane caused by repulsion between the hydrogen atoms, but it was studies on cyclohexane and its derivatives which emphasised the importance of *conformational analysis*.

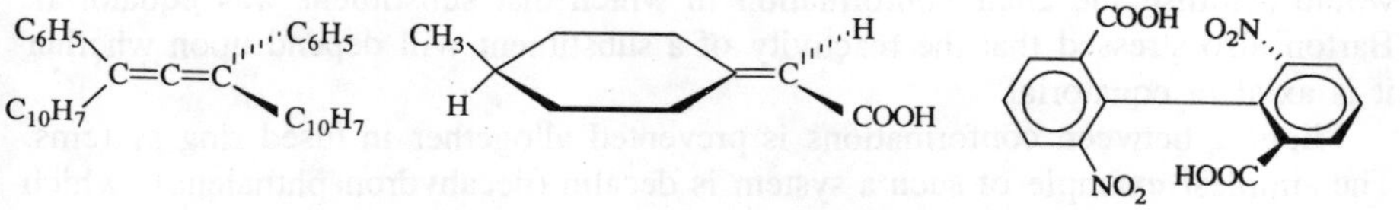

Figure 10.16 Some compounds exhibiting optical isomerism that do not contain asymmetric carbon atoms

In 1890 Ulrich Sachse had pointed out that the ring strains referred to by Baeyer in his strain theory would be eliminated altogether if six–membered (and larger) rings adopted a puckered arrangement. Molecular models showed that two arrangements (usually termed *boat* and *chair*) were possible for the cyclohexane molecule (Figure 10.17).

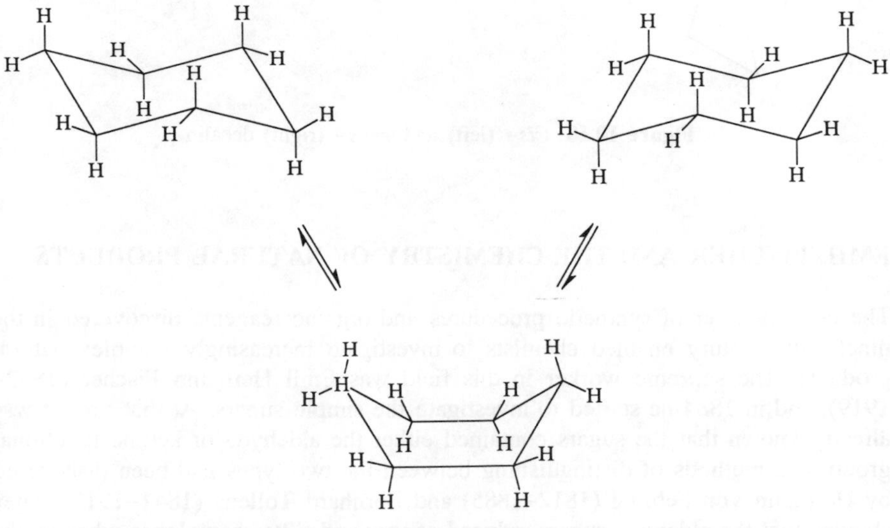

Figure 10.17 The two possible chair forms, and the boat form, of cyclohexane

In the 1930s, Odd Hassel (1897–1981) studied the geometry of cyclohexane derivatives by means of X–ray and electron diffraction studies. X–ray diffraction confirmed the existence of the chair form, and electron diffraction revealed that liquid cyclohexane was oscillating rapidly between the possible forms. In 1950 Derek H. R. Barton (*b.* 1918) stated the basic principles of conformational analysis as applied to cyclohexane derivatives. He pointed out that molecules with equatorial substituents would be generally more stable than those with axial substituents. Thus although cyclohexane itself has a small energy barrier to 'flipping' between the possible conformations, the presence of a bulky substituent would stabilise the chair conformation in which that substituent was equatorial. Barton also stressed that the reactivity of a substituent will depend upon whether it is axial or equatorial.

Flipping between conformations is prevented altogether in fused ring systems. The simplest example of such a system is decalin (decahydronaphthalene), which can exist as *cis* and *trans* isomers (Figure 10.18). These two isomers were first isolated by W. Hückel in 1925. The *trans*–decalin type of ring fusion occurs in many steroid molecules. The substituents are locked in either equatorial or axial positions, and this facilitated the elucidation of the relationship between reactivity and conformational position.

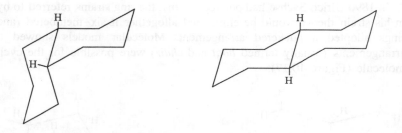

Figure 10.18 *Cis*– (left) and *trans*– (right) decalin

EMIL FISCHER AND THE CHEMISTRY OF NATURAL PRODUCTS

The large number of synthetic procedures and organic reagents discovered in the nineteenth century enabled chemists to investigate increasingly complex natural products. The supreme worker in this field was Emil Hermann Fischer (1852–1919), and in 1884 he started to investigate the simple sugars. At that time it was already known that the sugars contained either the aldehyde or ketone functional group, and methods of distinguishing between the two types had been discovered by Hermann von Fehling (1812–1885) and Bernhard Tollens (1841–1918). They showed that the aldehyde sugars reduced copper and silver complexes whereas the ketone sugars did not.

It was also known that the six–carbon sugars were polyhydroxy compounds and that they had the same molecular formula, $C_6H_{12}O_6$. Fischer proposed a general structure for the aldehyde sugars (Figure 10.20), and on the basis of van't Hoff's theory the four central atoms would all be asymmetric. Therefore there should be 2^4 or 16 isomers, existing as eight enantiomeric pairs.

Fischer (Figure 10.19) was a brilliant pupil at school, but his father persuaded him to enter the family timber firm rather than go to university. However, business was not to his taste, and his father declared: 'The boy is too stupid to become a businessman, so he had better become a student'. Fischer enrolled at the University of Bonn and at first seemed more interested in physics than chemistry. He later transferred to the University of Strasbourg to study with von Baeyer, and it was at Strasbourg that he was awarded his doctorate. It was while at Strasbourg that Fischer discovered phenylhydrazine, which he was to use so extensively in his later researches on sugars. In 1875 he moved to Munich with von Baeyer, and he subsequently held professorships at Erlangen, Würtzburg and Berlin. His work on carbohydrates and purines was widely acclaimed, and he was the second recipient of the Nobel Prize in Chemistry. However, his life had a tragic ending. Two of his three sons were killed in World War I, and his health was ruined by the toxic effects of mercury and phenylhydrazine. He contracted cancer and this, coupled with the depression caused by the death of his sons, led him to take his own life. His surviving son, Hermann Otto Fischer (1888–1960), had a distinguished career in organic chemistry and biochemistry.

Figure 10.19 Emil Hermann Fischer (1852–1919)

Fischer made extensive use of the reactions of the sugars with phenylhydrazine, a reagent which he had discovered in 1875. Fischer also oxidised the sugars to mono– and dicarboxylic acids, reduced the carbonyl function to a hydroxyl group, and prepared lactones (cyclic esters) from the monocarboxylic acids. Fischer also used a reaction devised by Heinrich Kiliani (1855–1945) in which the carbon chain of a sugar could be extended. Thus a five–carbon sugar could be converted to two six–carbon sugars because a new asymmetric centre had been added.

By 1891 Fischer had established the open chain structure of glucose and some of the other simple sugars. He continued his studies for several more years, establishing the structures of all eight possible pairs of aldohexoses, including gulose and idose, which are not naturally occurring.

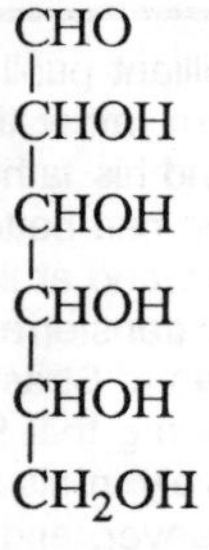

$$
\begin{array}{c}
\text{CHO} \\
|\\
\text{CHOH} \\
|\\
\text{CHOH} \\
|\\
\text{CHOH} \\
|\\
\text{CHOH} \\
|\\
\text{CH}_2\text{OH}
\end{array}
$$

Figure 10.20 The general structure for the six–carbon aldehyde sugars

Problems remained with the glucose structure, in that it provided no explanation for the phenomenon of mutarotation or for the rather feeble aldehydic properties of glucose. In 1893 Fischer prepared a methyl glucoside (a methyl ether of glucose), and the following year W. Alberda van Eckenstein (1858–1937) prepared another, and therefore it seemed that the carbonyl carbon had been changed into another asymmetric centre. Cyclic structures were proposed for the methyl glucosides, but Fischer felt that there was insufficient evidence to propose a cyclic structure for glucose itself. Around 1926 Haworth and Edmund L. Hirst established that the methyl glucosides occur most commonly as six–membered rings, although Haworth later established that a five–membered ring is also possible. It became clear that glucose itself exists as two six–membered ring forms (α– and ß–glucose) in equilibrium with the open chain aldehyde (Figure 10.21).

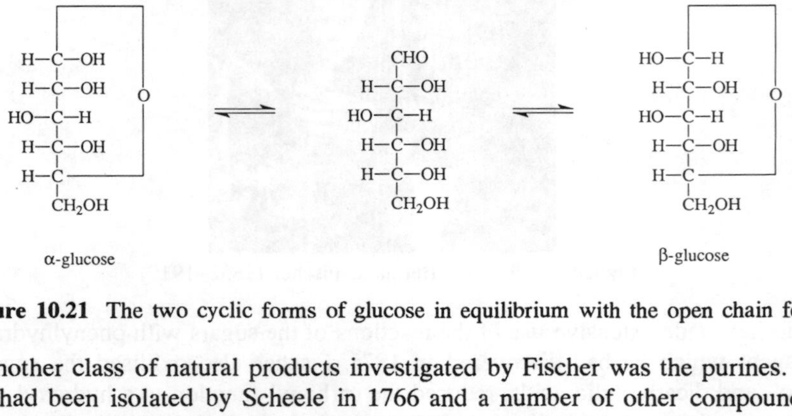

Figure 10.21 The two cyclic forms of glucose in equilibrium with the open chain form

Another class of natural products investigated by Fischer was the purines. Uric acid had been isolated by Scheele in 1766 and a number of other compounds in the group had been obtained before Fischer began studying caffeine in 1881. Among the other members of the group studied by Fischer were xanthine, theobromine and guanine. Fischer synthesised some of the purines and established the relationship between them.

Fischer also performed important work on proteins. During the nineteenth century a number of amino acids had been isolated and their structures established. The highly complex nature of the proteins themselves was not realised at first; Liebig and his contemporaries attempted to assign simple molecular formulae to them on the basis of elemental analyses. However, it gradually became clear that proteins consisted of amino acids joined together. Fischer suggested that the amino acids were linked via the grouping –CONH– and he proceeded to join amino acids by forming this linkage. He proposed the name *polypeptides* for these synthetic materials. Fischer succeeded in synthesising a polypeptide containing eighteen amino acid residues. He found that this and other peptides behaved like the intermediate products of protein hydrolysis (which at that time were called *peptones*).

Natural product chemistry has continued to make rapid advances, although no individual has made a contribution to rival that of Emil Fischer. Carbohydrate chemists have concentrated on molecules in which individual sugar molecules are linked together, and have now solved the structures of many oligo– and polysaccharides. Other classes of compound which have been studied include the steroids and terpenes. Extensive work has also been done on groups such as the vitamins and the hormones, in which the individual members belong to different chemical classes.

Although Fischer had convincingly demonstrated the nature of the peptide link, the determination of the amino acid sequence of naturally occurring peptides and proteins had yet to be accomplished. A crucial contribution in that area was made in 1945 by Frederick Sanger (*b.* 1918), who introduced the reagent 2,4–dinitro-fluorobenzene, which reacts with the free amino group of the N–terminal amino acid of a protein to produce the dinitrophenyl (DNP) derivative. After hydrolysis of the protein, the DNP amino acid (which is coloured yellow) is separated and identified by partition chromatography. Another important development was the degradation method introduced by P. Edman in 1950. This utilises the reagent phenyl isothiocyanate to cleave the N–terminal residue from a peptide. The cleaved amino acid derivative is then identified, and the procedure repeated to identify the next residue, and so on. To determine the amino acid sequence of a large protein, it is first partially hydrolysed into smaller peptides by chemical and enzymic means before the chemical degradative processes are used. By the application of these techniques, Sanger established the complete amino acid sequence of the hormone insulin, and the sequences of many other proteins have now been determined.

Fischer's discoveries in the field of natural products were made solely by the application of the techniques of organic chemistry. More recent investigations have relied to an increasing extent on physical methods, both to separate and purify the substances under investigation and to determine their structures. Natural product chemistry benefited enormously from the general introduction of chromatographic separation techniques in the 1930s, and something of a turning point in structure determination was reached in 1944 when Dorothy Hodgkin determined the structure of penicillin by X–ray diffraction before the organic chemists had completed their investigations. More recently, nuclear magnetic resonance and mass spectroscopy have been used in structural studies (Chapter 13).

ORGANIC SYNTHESIS IN THE TWENTIETH CENTURY

At the beginning of the twentieth century, it was generally accepted that the synthesis of a natural product provided the final and essential confirmation of its

structure. Such is the power of modern physical methods that this view is no longer
held. There have nevertheless been some heroic syntheses in recent years. One of
the most outstanding achievements since World War II was the 1971 synthesis of
vitamin B_{12} ($C_{63}H_{90}CoN_{14}O_{14}P$) as a result of a collaborative effort between teams
led by Robert Burns Woodward (1917–1979) at Harvard and Albert Eschenmoser
in Zürich. This project involved 100 chemists and took eleven years.

Woodward (Figure 10.22) proved to be an excellent student and he went to
the Massachusetts Institute of Technology when he was sixteen. He was
awarded his PhD when he was twenty and he spent the remainder of his
professional life at Harvard. He was the most influential organic chemist of his
generation. He was quick to realise the power of the new spectroscopic
techniques in organic chemistry, and his first important work was the
correlation of the ultraviolet absorption spectra of α,β–unsaturated ketones
with their structure. He deduced the structure of several important natural
products, including penicillin and strychnine. He is, however, chiefly
remembered for his brilliant contributions to organic synthesis. One of his
early successes was the synthesis of quinine, which had been the original
object of Perkin's work in 1856 when he discovered the first synthetic organic
dye (Chapter 15). Among Woodward's many other syntheses were cholesterol
in 1951, chlorophyll in 1960 and the tetracycline antibiotics in 1962. His most
famous achievement was the synthesis, in collaboration with Eschenmoser, of
vitamin B_{12} in 1971. Many new reactions and techniques were developed
during the course of Woodward's synthetic studies.

Figure 10.22   Robert Burns Woodward (1917–1979)

Although synthesis is now less frequently used to confirm the structure of a
natural product, novel compounds are being prepared at an unprecedented rate.
Many new classes of compound have been discovered, and thousands of substances
are prepared annually for possible use as drugs, agrochemicals, plastics and so on.
Many novel synthetic procedures have been developed, of which we can mention
only a few.

A remarkable ring–building reaction was announced in 1928 by Otto Diels
(1876–1954) and Kurt Alder (1902–1958). Their research commenced with a study

of the addition reactions of esters of azodicarboxylic acid, $RO_2C-N{=}N-CO_2R$. They discovered that these compounds added to alkylated butadienes (e.g. isoprene). Further investigation revealed that this reaction was not restricted to azo compounds, and that unsaturated hydrocarbons and their derivatives would also act as *dienophiles* (Figure 10.23). The *Diels–Alder reaction* has been extensively employed in organic synthesis, and its discoverers were awarded the Nobel Prize in Chemistry in 1950.

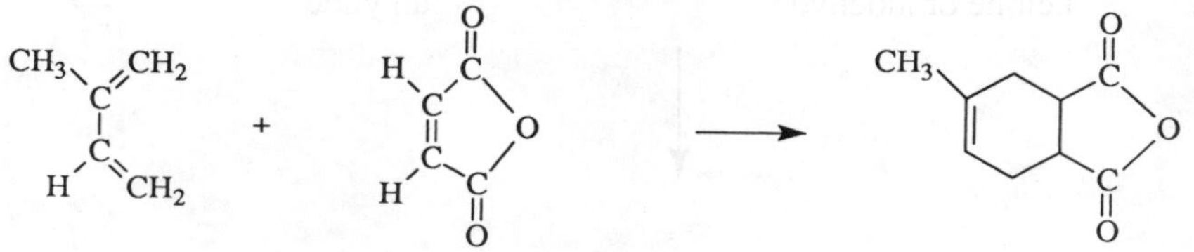

Figure 10.23 The Diels–Alder reaction between isoprene (the diene) and maleic anhydride (the dienophile) to yield 4–methyl–1,2,3,6–tetrahydrophthalic anhydride

$$(C_6H_5)_3\overset{\oplus}{P}{-}\overset{\ominus}{C}H_2$$

Figure 10.24 Methylenetriphenylphosphorane (an *ylide*)

Another notable development has been the introduction of reagents containing boron or phosphorus. In 1953 Georg Friedrich Karl Wittig (1897–1987) introduced a synthetic procedure utilising organophosphorus compounds called *ylides* (Figure 10.24). He had become interested in these compounds while attempting to join five organic groups to a phosphorus atom. He found that the ylides reacted smoothly with carbonyl compounds to yield an alkene (Figure 10.25). This reaction has proved to be very valuable, and is used in the commercial production of ß–carotene and many other compounds.

In 1956 Herbert C. Brown (*b.* 1912) discovered that in ethereal solution diborane (B_2H_6) dissociates into borane (BH_3), which can add to an alkene. The organoborane formed can be converted to an alcohol by treatment with hydrogen peroxide, and the overall result of this *hydroboration* reaction is the anti-Markovnikoff addition to the double bond. Brown was also responsible for the introduction into organic chemistry of the reducing agents sodium borohydride (sodium tetrahydridoborate(III)) and lithium aluminium hydride (lithium tetrahydridoaluminate(III)).

After Fischer's synthesis of an octadecapeptide in 1907, peptide synthesis has made steady progress with the development of protecting groups for reactive side chains and improved coupling procedures. A novel approach was pioneered in 1962 by Robert Bruce Merrifield (*b.* 1921) in which an amino acid is anchored to an insoluble polymeric support and successive amino acids are then coupled to it. When completed, the peptide is cleaved from the support. The technique, known as *solid–phase synthesis*, has been automated and has also been adapted to the preparation of nucleic acids.

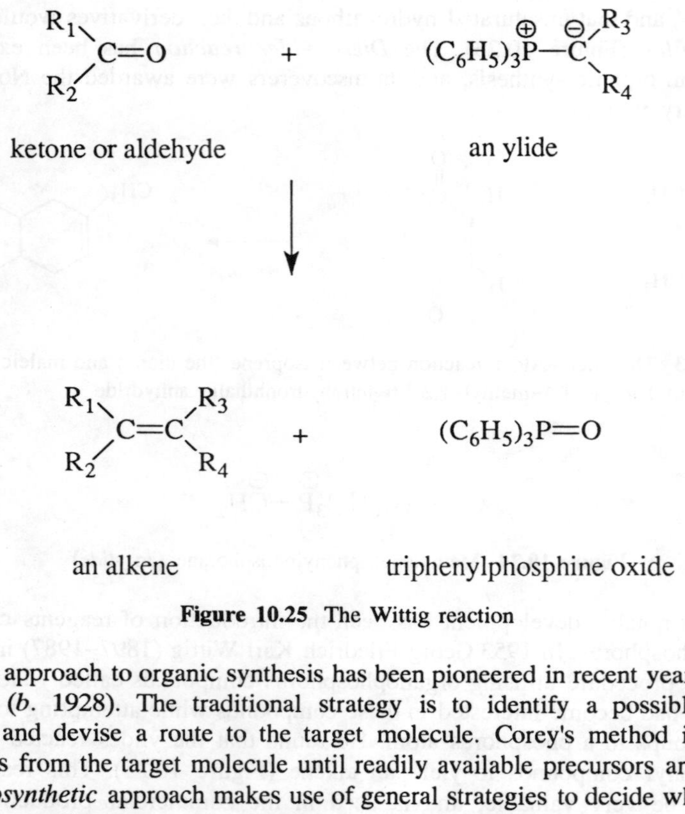

$$R_1 \atop R_2 {>} C{=}O \qquad + \qquad (C_6H_5)_3 \overset{\oplus}{P}{-}\overset{\ominus}{C} {<} {R_3 \atop R_4}$$

ketone or aldehyde   an ylide

$$R_1 \atop R_2 {>} C{=}C {<} {R_3 \atop R_4} \qquad + \qquad (C_6H_5)_3P{=}O$$

an alkene   triphenylphosphine oxide

Figure 10.25 The Wittig reaction

A new approach to organic synthesis has been pioneered in recent years by Elias J. Corey (*b*. 1928). The traditional strategy is to identify a possible starting structure and devise a route to the target molecule. Corey's method is to work backwards from the target molecule until readily available precursors are reached. This *retrosynthetic* approach makes use of general strategies to decide which of the possible immediate precursors to a compound in a synthetic sequence is likely to form part of the best overall route. This method has been greatly enhanced by the use of computers. With the advent of computer graphics, it is now possible to input a chemical structure and to display it on screen. The computer program has a database of organic reactions, and the chemist is able to devise the route to the target molecule retrosynthetically on screen.

Advances in organic chemistry since the 1930s have been assisted by the application of electronic theories of chemical bonding and by the elucidation of reaction mechanisms (Chapter 13). Since 1950 it has been realised that many reactions proceed via mechanisms which do not involve the attack of electrophilic or nucleophilic reagents, but involve *free radicals*. We have seen in Chapter 8 how Bunsen, Kolbe and Frankland attempted to prepare free radicals, but after Cannizzaro's systematisation of atomic and molecular weights in 1860 it was realised that all the so–called free radicals prepared up till that time were in fact dimers. Then in 1900 Moses Gomberg (1866–1947), while attempting to prepare hexaphenylethane, obtained the highly reactive triphenylmethyl radical. We now know that this species, like other free radicals, possesses an unpaired electron. Free radicals may be generated photochemically, or by the decomposition of an unstable compound such as dibenzoyl peroxide, and may initiate *chain reactions* in which many individual reactions ensue from the initial attack of one free radical.

11

Atomic Structure, Radiochemistry and Chemical Bonding

In previous chapters we have seen how Dalton's atomic theory encountered difficulties, especially during the first half of the nineteenth century. After 1860 the chemical atom became so useful, especially to organic chemists, that only a few dissenting voices were heard in this branch of the science. However, to many physicists the atomic concept seemed unnecessary, and some of the great advances in nineteenth century physics had been made without reference to atomism. Then, in a series of discoveries which were made in rapid succession around the turn of the century, the existence of atoms was established to the satisfaction of everyone. All atoms were shown to contain identical subatomic particles, which were called *electrons*. Furthermore, certain atoms were shown to be undergoing spontaneous and continuous transmutation into others. The atomic concept may have been vindicated, but at the cost of disproving Dalton's tenets of the indivisibility and immutability of atoms. Further advances placed the periodic table on a firm theoretical basis, and provided an explanation of the forces involved in chemical bonding.

CATHODE RAYS

The early ideas on atomic structure followed from studies on the passage of electricity through rarefied gases. As early as 1821, Davy had observed a luminosity accompanying the passage of a spark, and he noticed that the colour depended on the residual gas in the tube. In the middle of the century Heinrich Geissler (1814–1879) devised the mercury vapour pump, and with its aid was able to manufacture tubes in which the gaseous pressure was much lower than anything previously available. Using Geissler tubes, Julius Plücker (1801–1868) noticed that the phosphorescent spot which appeared on the glass when the tube was in operation could be moved by a magnet, thus suggesting that the beam emanating from the cathode was electrically charged. Eugen Goldstein (1850–1930) observed that objects placed in the beam cast a shadow on the glass wall of the tube. He coined the term *cathode rays* to describe the radiation.

These experiments were repeated and extended in 1879 by William Crookes, the discoverer of thallium (Chapter 9). Crookes demonstrated that the luminous effects observed by Davy and others were distinct from the phenomena observed in very highly evacuated tubes, when the luminosity diminishes but the phosphorescence of the glass increases. Crookes believed that in the very highly evacuated tubes in a state of discharge there appeared a fourth state of matter, which he termed *radiant matter*. He demonstrated that the beam of radiant matter was deflected by a magnet in a direction indicating a negative charge, and he showed that a small paddle wheel placed in the beam was made to rotate (Figure 11.1). He interpreted this as being evidence that radiant matter is particulate in nature, and he called the particles *negatively electrified molecules*. Concluding his account of his work, Crookes said: '... here, it seems to me, lie Ultimate Realities, subtle, far reaching, wonderful'.

Figure 11.1 Crookes's paddle wheel apparatus. Crookes thought that the rotation of the paddle wheel was due to the impact of particles. It is now known that the rotation is caused by one side of a vane becoming warmer as a result of electron bombardment, and the molecules of the residual gas in the tube recoiling with greater energy when they collide with the heated surface

THE ELECTRON

Towards the end of the nineteenth century there were two distinct theories concerning the nature of cathode rays. One school of thought, to which most German physicists subscribed, was that cathode rays were waves moving through the *ether*. This theory likened cathode rays to light, which was thought to be propagated in this way. The other view was the negative–particle theory of Crookes. In 1897 Joseph John Thomson (1856–1940) published the results of some experiments which he had conducted to test the particle theory.

Supporters of the wave theory had argued that the negative particles ejected from the cathode were distinct from the rays which caused the phosphorescence on the glass. Thomson built a tube containing a collecting cylinder connected to an electrometer. When the phosphorescent spot was deflected by a magnet into the cylinder, the electrometer detected a charge in the cylinder.

Having established that the electrical and phosphorescent effects of cathode rays could not be separated, Thomson proceeded to demolish another objection to the charged–particle theory. This was that no deflection of the cathode ray beam had been observed when it was passed between two plates maintained at a small potential difference. Thomson argued that the passage of the rays rendered the residual gas in the tube conducting, thus annulling the charge on the plates. Thomson showed that a deflection did in fact occur under conditions of very high vacuum, but that there was no deflection when the vacuum was not so good. Thomson remarked:

'... I can see no escape from the conclusion that they [the cathode rays] are charges of negative electricity carried by matter. The question arises, What are these particles? Are they atoms, or molecules, or matter in a still finer state of subdivision? To throw some light on this point I have made a series of measurements of the ratio of the mass of these particles to the charge carried ...'.

Thomson (Figure 11.2) was born at Manchester where his father was a bookseller. His ambition was to become an engineer, but the death of his father meant that his family could not afford the extra fees which were charged to students of this subject. Instead, he concentrated on the study of mathematics, physics and chemistry, and won a scholarship to Trinity College, Cambridge. As a result of an outstanding examination performance he became a fellow of Trinity in 1880, and he remained at the college until his death 60 years later. His early research was theoretical, and many were surprised when in 1884 he was elected to the Cavendish Professorship in Experimental Physics. He proved to have a genius for devising experiments, although he was not a particularly skilful laboratory worker. He became Master of Trinity College in 1918, when he relinquished the Cavendish Professorship. He is remembered not only because his work was of the greatest significance, but also because under his leadership the Cavendish Laboratory in Cambridge became a world famous centre for research in atomic physics.

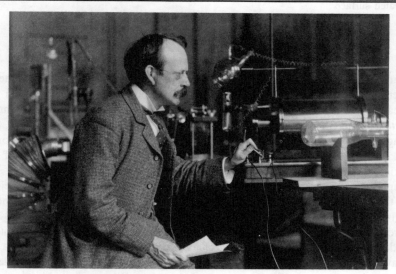

Figure 11.2 Joseph John Thomson (1856–1940)

Thomson measured the extent of the deflection produced in the beam by a given electric field, and he then found the strength of the magnetic field necessary to produce the identical deflection (Figure 11.3). The value Thomson calculated for the mass–to–charge ratio (m/e) was approximately one–thousandth of that for the hydrogen ion, and was independent of the residual gas in the tube. Thomson devised some ingenious experiments to measure the value of the charge carried by the new particle, and found that it was of the same order of magnitude as that of the hydrogen ion. This indicated that the particle, which was given the name *electron*, possessed a very small mass.

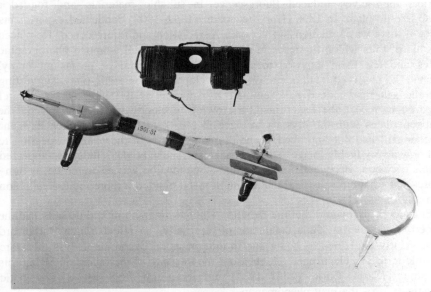

Figure 11.3 The apparatus used by Thomson to measure the mass–to–charge ratio of the electron

Between 1908 and 1917 the American physicist Robert Andrews Millikan (1868–1953) performed an elegant series of experiments to obtain an accurate value of the charge carried by the electron (e). Millikan measured the rate of rise of a small charged oil droplet of known mass in an electric field, and was thereby able to calculate the charge on the droplet. In all cases this was found to be an integral multiple of a constant value, which therefore represented the magnitude of a unit charge. Millikan's value was very close to the currently accepted figure of 1.602×10^{-19} C. The values obtained by Thomson and Millikan, in conjunction with Thomson's mass/charge determinations, pointed to a very small mass for the electron. Current estimates accord the electron a mass of 1/1837 on the relative atomic mass scale.

Not surprisingly, Thomson's suggestion of the existence of subatomic particles common to all atoms was received with some scepticism, but in 1904 he proposed a model for the structure of the atom to fit the available evidence. He suggested that the atom was a sphere of uniform positive charge within which rings of electrons may rotate. The mass of the atom was almost entirely accounted for by the mass of the electrons, and on this basis the hydrogen atom would contain a very large number of electrons.

RADIOACTIVITY

In 1895 Wilhelm Konrad Röntgen had discovered that the fluorescent spot on a discharge tube emitted a very penetrating radiation, which he called X–radiation. X–rays themselves induced fluorescence in suitable materials and fogged a photographic plate. In 1896 Henri Becquerel (1852–1908) conducted experiments to see if there was a relationship between X–radiation and fluorescence. He placed crystals of salts which fluoresce in ultraviolet light on well wrapped photographic plates and left them exposed to sunlight. With a crystal of potassium uranyl sulphate he observed a silhouette on the developed photographic plate, suggesting that the salt may indeed have emitted X–radiation as it fluoresced. Becquerel prepared to repeat the experiment, but owing to cloudy weather he placed the wrapped plates with the crystals in place in a dark drawer. After three days the weather still had not improved, so he developed some of the plates, expecting to find only a very feeble image. To his amazement, much bolder silhouettes appeared on this occasion. Becquerel found that the new radiation was also emitted by non–fluorescent uranium salts, and that uranium metal emitted a more intense radiation than any of its salts.

Becquerel discovered that the radiation emitted by uranium and its salts had the property of rendering air a conductor of electricity, an effect which Rutherford subsequently explained as being due to the ionisation of the air. It was therefore possible to use an electrometer to measure the intensity of the radiation. Using this technique Marie Curie (1867–1934) embarked in 1898 on a systematic investigation of the new phenomenon.

Curie examined all the known elements and found that, apart from uranium, only thorium appeared to emit ionising radiation. All compounds of these elements were active, and changing the temperature had no effect on the activity. The new property, which Curie named *radioactivity*, appeared to be an atomic phenomenon, because the intensity of the radiation emitted by a uranium compound was found to be proportional to the percentage of uranium it contained.

Curie measured the intensity of the radiation from the mineral pitchblende. The principal metal in pitchblende is uranium, but smaller quantities of many other metals are present as well. Curie found that the radioactivity of pitchblende was considerably greater than could be explained on the basis of its uranium content, and she concluded that there was an unknown and highly radioactive element present. She dissolved some pitchblende in acid and precipitated each metal in turn and checked with the electrometer to see if the radioactivity had come out of solution. She found that a high radioactivity was associated with the bismuth fraction, and named the new element which was clearly present *polonium* in honour of her native country. However, the spectrum of the precipitate contained no new lines, so it appeared that polonium must be present in extremely small quantity.

Curie also found that the barium fraction was radioactive, and on careful crystallisation of the chloride she found that the least soluble portion was 900 times more radioactive than uranium. The salt still contained large quantities of barium chloride, but it glowed in the dark, and Curie christened the new element present *radium*. The spectroscopist Eugène Demarçay observed new lines in the spectrum of the salt, thus confirming the existence of a new element. It was clear that a huge quantity of pitchblende would be required to obtain a sensible quantity of radium or one of its compounds. Curie obtained one ton of pitchblende residues (from which the uranium had been extracted) as a gift from the Austrian government. After four years of hard work, Curie and her husband Pierre succeeded in isolating 0.1 g of pure radium chloride.

Figure 11.4 Marie Curie (1867–1934) and her husband Pierre in their laboratory

A third new radioactive element was discovered in the pitchblende residues by André Debierne (1874–1949), who was working with the Curies. It was precipitated with the rare–earth elements, but like polonium it was present in such minute quantity that neither the element nor any of its compounds could be isolated in a pure form. This element was named *actinium*.

Marie Curie (Figure 11.4) was born Marya Sklodovska, the daughter of a physics teacher in Warsaw. Her family could not at first afford to send her to university, so between the ages of 18 and 24 she worked as a governess. She then studied physics and mathematics at the Sorbonne in Paris, and shortly after graduating she married the physicist Pierre Curie. Her work on radioactivity was performed for her doctorate, and her thesis has been described as one of the most important ever submitted for such a degree. When the significance of Marie's work became apparent, Pierre abandoned his own research and thereafter they worked together. Their laboratory was an abandoned shed, and their isolation of a minute quantity of pure radium chloride from a huge amount of pitchblende represents one of the most heroic struggles in the history of science. In 1903 the Curies and Becquerel were awarded the Nobel Prize for Physics. In 1904 Pierre became professor of physics at the Sorbonne. He was killed in 1906 by a wagon as he crossed the street, and Marie then succeeded to his professorship. In 1911 she was awarded the Nobel Prize for Chemistry, thus becoming the first person to win two Nobel prizes. During World War I she organised an X–ray service to locate bullets and shrapnel in injured soldiers. After the war she travelled to the United States, where she was presented with one gram of pure radium and a sum of money to enable her to continue her research. She died of leukaemia, contracted as a result of prolonged exposure to radiation.

ALPHA, BETA AND GAMMA RADIATIONS

While the Curies were investigating new radioactive elements, others were studying the emissions from radioactive substances. Foremost in this field was Ernest Rutherford (1871–1937). Working at the Cavendish Laboratory in Cambridge in 1898, he demonstrated that there were two kinds of radiation, which he called *alpha* and *beta rays*. The beta rays were about as penetrating as X–rays, but most alpha rays were stopped by thin sheets of metal foil. In 1899 Becquerel showed that the beta rays could be deflected in a magnetic field in a direction showing that they carried a negative charge. It seemed likely that beta rays were in fact electrons, and, using a method similar to Thomson's, Becquerel measured their mass/charge ratio and obtained a value similar to that which Thomson had found for the electron.

In 1903 Crookes devised his *spinthariscope* in which alpha rays fell on a zinc sulphide screen. Tiny flashes of light were observed, thus showing that the alpha ray was a stream of particles. Alpha rays were only deflected very slightly by a magnetic field in a direction which indicated a positive charge. In 1903 Rutherford succeeded in measuring this deflection and used it to calculate the mass/charge ratio. His first value was roughly similar to that of the hydrogen ion, but a more accurate determination in 1906 yielded a value of approximately double that of the hydrogen ion. Once the particulate nature of alpha and beta radiations had been demonstrated, the names alpha and beta *particle* slowly came into use.

A third radiation was observed in 1903 by P. Villard. Since it was highly penetrating and undeflected by a magnetic field, Rutherford suggested that it was electromagnetic radiation of shorter wavelength than X–rays, and he named the emission *gamma radiation*.

Figure 11.5 Ernest Rutherford (1871–1937)

Rutherford realised that his result for mass/charge for the alpha particle was consistent with it being a doubly positively charged helium ion. Terrestrial helium had recently been obtained by Ramsay (Chapter 9) by heating the uranium mineral cleveite, and it was soon found that other uranium minerals produced the gas. Direct experimental proof of the nature of the alpha particle was obtained in 1909 by Rutherford and Royds, who allowed alpha particles from *emanation*, a radioactive gas emitted by radium, to pass through a glass tube whose walls were only 0.01 mm thick. The alpha particles were collected in a thicker outer glass tube, and after six days the contents of this tube were compressed and the spectrum recorded. The lines of helium were clearly visible. Rutherford and Royds ended their paper with the words: 'We can conclude with certainty from these experiments that the alpha particle after losing its charge is a helium atom'.

Rutherford (Figure 11.5) was born on a small farm near Nelson in New Zealand. He won first class honours at Canterbury College in Christchurch, and then commenced research on electromagnetism. In 1895 he won a scholarship to travel to the Cavendish Laboratory in Cambridge to work with J. J. Thomson. He continued his work on electromagnetism for a further year, and then he and Thomson investigated the conductivity induced in air by the X-radiation recently discovered by Röntgen. Rutherford realised that the conductivity was due to the ionisation of the air, and he soon showed that radioactive emissions produced a similar effect. In 1898 he moved to McGill University in Montreal, where he formed his productive partnership with Soddy. He returned to England in 1907 to become professor at Manchester. During World War I he served on the War Research Committee. One day in 1917 he arrived late for a meeting and by way of explanation he said: 'I have been engaged on experiments which suggest that the atom can be artificially disintegrated. If it is true, it is of far greater importance than a war'. In 1919 he succeeded Thomson as professor at the Cavendish in Cambridge, where he more than maintained the pre-eminent position of that laboratory in the realm of atomic physics. His great genius lay in devising crucial experiments whose results were of the utmost significance but which employed relatively simple apparatus. He continued to head the Cavendish until his death.

THE MECHANISM OF THE RADIOACTIVE PROCESS

Evidence had also been accumulating which enabled a theory to be advanced concerning the mechanism of the radioactive process. An important experiment was performed by Crookes in 1900. He found that he was able to obtain a radioactive precipitate from a solution of a uranium salt. The uranium, which remained in solution, had become almost totally inactive. This was a remarkable finding, as the recent work of the Curies had indicated that the activity was a property of the uranium atom. Crookes called the active substance in the precipitate *uranium X*. After a year he re-examined the uranium X and found that it was now inactive, while the original uranium had regained its activity. In retrospect, it is fortunate that Crookes used a photographic method to measure the activity. The alpha particles, which uranium continues to emit, would not penetrate the wrapping of the photographic plate and were not detected.

Becquerel obtained similar results with a uranium solution using a different precipitating reagent, and in 1902 Rutherford, working with Frederick Soddy (1877–1956) (Figure 11.6), precipitated radioactive *thorium X* from a solution of a thorium salt. Over a period of a month thorium X became inactive while the original thorium in solution regained its activity. In 1900 Friedrich Ernst Dorn had observed that a radioactive gas was given off by radium, and in the same year Owens and Rutherford found that a similar gas was produced by thorium. Both these *emanations* were alpha emitters; the radioactivity of the gas from thorium lasted for only a few minutes, but that from radium lasted for a few days. Surfaces which had been in contact with these emanations were found to be contaminated with radioactive deposits. Ramsay and Whytlaw–Gray measured the atomic weight of radium emanation and found it to be 222, approximately 4 units less than that of radium.

Figure 11.6 Frederick Soddy (1877–1956)

On the basis of this evidence, Rutherford and Soddy in 1902 put forward their disintegration hypothesis. They proposed that radioactive elements were undergoing spontaneous transformation into new elements, and therefore the atoms of radioactive elements were breaking down into new atoms. Within the space of five years the Daltonian concept of the chemical atom had undergone a radical change. Not only had the existence of subatomic particles (electrons) been demonstrated, but also transmutation was shown to be a natural phenomenon. In their paper of 1902 Rutherford and Soddy also demonstrated that the decay of a radioactive substance followed an exponential law.

Once it became clear that radioactive elements were decaying to new elements, which were themselves radioactive, a great deal of effort was expended in working out the decay sequences. In most cases the new elements were at first obtained in quantities too small to be weighed and were distinguished from each other only by the type of decay they exhibited and the rate at which the decay occurred. Three *decay series* were elucidated: the uranium series proceeded through radium and terminated with the stable radium G; the thorium series ended in the stable thorium D; and the actinium series ended in actinium D. Between them, these series contained around 25 new radioelements.

ISOTOPES

Although these new elements were obtained in quantities which were far too small to enable their atomic weights to be determined, attempts were made to assign them to positions in the periodic table on the basis of their chemical properties.

These investigations uncovered the problem that some of the new elements were so similar to familiar elements that once mixed with them they could not be separated again. Thus in 1906 Bertram Borden Boltwood (1870–1927) of Yale University was unable to separate ionium (the immediate precursor of radium in the uranium series) from thorium. Similarly, Geörgy Hevesy (1885–1966) was unsuccessful in his attempts to separate radium D from lead.

The solution to these problems was proposed simultaneously and independently by Kasimir Fajans (1887–1975) and Frederick Soddy in 1913. Their conclusions, now known as the *group displacement laws*, were that when an alpha particle is emitted the new element occupies a position two places to the left in the periodic table, whereas the loss of a beta particle produces an element one place to the right of the parent. Following from this was the revolutionary proposal that the elements between lead and uranium in the periodic table could exist as more than one kind of atom, differing in mass but displaying identical chemical properties. The problems encountered by Boltwood in his attempts to separate ionium from thorium were now explained on the grounds that ionium *was* thorium. Fajans called a group of chemically identical atoms with differing atomic weights a *pleiad*, but Soddy's term *isotopes* has been generally adopted. It was therefore proposed that most of the new radioelements were isotopes of existing elements. The gaseous emanations were shown to be isotopes of the same new rare–gas element, now known as *radon*.

Figure 11.7 Theodore William Richards (1868–1928)

The group displacement laws predicted that the end–products of the three decay series were stable isotopes of lead with differing atomic weights. It therefore followed that samples of lead of different mineral origin should have different isotopic compositions and hence different atomic weights. The acknowledged expert of the day on atomic weight determination was Theodore William Richards (1868–1928) (Figure 11.7) of Harvard University. He had brought new standards of accuracy to the traditional gravimetric methods. Fajans sent Max Lembert (1891–1925) from Karlsruhe to Harvard to learn Richards's techniques and to determine the atomic weight of lead from different sources. Richards and Lembert published their results in 1914, and, as they themselves remarked, the outcome was striking. Common lead was found to have an atomic weight of 207.15, whereas that from North Carolina uraninite gave a value of 206.40. The differences between these values was many times the experimental error for the technique. Other workers measured the atomic weight of lead from other sources, and the highest value (207.90) was obtained by Hönigschmid for lead from a thorite mineral.

Figure 11.8 Francis William Aston (1877–1945) with his mass spectrograph

At the time that Soddy and Fajans were postulating that heavy elements exist as isotopes, Thomson was obtaining evidence which was to point to a similar conclusion for neon. Thomson had conducted a long series of experiments on the *positive rays* which pass through a perforated cathode in a discharge tube. It became clear that these were composed of positive ions derived from the residual gas in the tube. By deflecting the rays with electric and magnetic fields arranged at right–angles, ions with similar mass/charge values fell on a parabola at the end of the tube, which acted as a fluorescent screen. By 1913 Thomson's apparatus was sufficiently refined for him to estimate the masses of the positive ions (assuming they carried unit positive charge). He found that neon gave two parabolas, corresponding to atomic weights of 20 and 22. Thomson initially thought that the heavier ion might be derived from the compound NeH_2 formed between neon and traces of hydrogen in the tube. His assistant, Francis William Aston (1877–1945), suspected that the explanation might reside in the existence of two different types of neon atom, and he attempted to separate them by repeated diffusion through an apparatus made of pipe–clay, but was unsuccessful.

This work was halted by World War I, but in 1919 Aston resumed his investigations. He recorded the positive ions in the discharge tube by arranging for them to fall on a photographic plate, and he called his instrument the *mass spectrograph* (Figure 11.8). He demonstrated conclusively that both the lines Thomson had observed were due to neon alone, and he also showed that there was a third isotope of neon of very low abundance with atomic weight 21. Aston developed the apparatus still further and soon demonstrated that many other elements occur as isotopes. Aston measured his atomic masses on the scale $^{16}O =$ 16.0000, and in his *whole–number rule* stated that isotopic masses were integral on this scale. Fractional atomic weights were due to a mixture of isotopes present in the natural element.

Mass spectrographs were also built in the United States by A. J. Dempster and K. T. Bainbridge. The mass spectrograph has been succeeded by the *mass spectrometer,* in which the intensity of the separated ion beams are measured electrically. These instruments are now widely used in the determination of molecular structure (Chapter 13). The term *relative atomic mass* is now used in place of atomic weight, and isotopic masses are measured on the $^{12}C = 12.0000$ scale. Aston himself soon discovered that small deviations from the whole–number rule are the norm.

THE NUCLEAR ATOM

Soon after Thomson published his theory that the atom was a sphere of uniform positive charge containing rotating rings of electrons, experimental evidence started to accumulate which was inconsistent with it. When tracks of alpha particles were photographed by means of the cloud chamber, it was observed that they sometimes made an abrupt turn. In 1906 Rutherford observed that a beam of alpha particles would penetrate very thin metal foil, but a photographic plate placed behind the foil registered a diffuse spot, indicating that some of the particles had been deflected. The phenomenon was studied carefully by Hans Geiger (1882–1945) and Ernest Marsden, who found that, although most particles experienced either no deflection or a very small one, a small minority were sharply deflected, some through angles greater than 90°. As Rutherford later remarked: 'It was about as credible as if you had fired a 15 inch shell at a piece of tissue paper and it had come back and hit you'.

On the basis of these results, Rutherford in 1911 postulated that the atom consists of a tiny central positively charged region, which he subsequently termed the *nucleus*. The nuclear positive charge was balanced by electrons revolving round the nucleus at a considerable distance. The results of the alpha particle scattering experiments were thereby explained; the positive alpha particle would experience little or no deviation unless it happened to approach very close to the positively charged nucleus.

NUCLEAR CHARGE AND ATOMIC NUMBER

Attention was then directed to the question of the magnitude of the nuclear positive charge. In 1913 Geiger and Marsden re–examined the scattering of alpha particles by gold foil and concluded that 'the number of elementary charges composing the centre of the atom is equal to half the atomic weight'. A similar conclusion had been reached two years earlier by Charles Barkla (1877–1944) for light elements up to atomic weight 32 by studying the energies of X–rays scattered by those elements.

In 1913 the Dutchman Antonius Johannes van den Broek (1870–1926) proposed that since the average difference in atomic weight between successive elements in the periodic table was about two units, the number of nuclear charges would differ by one unit between adjacent elements. He suggested that the nuclear charge number defined an element's position in the periodic table. However, he thought that the charge on the uranium nucleus was 118, and on this basis concluded that many elements remained undiscovered.

Direct experimental evidence on the magnitude of nuclear charge was provided in 1913 by Henry Moséley (1887–1915) (Figure 11.9) as a result of studies of the X–ray spectra of elements. Since the discovery of X–rays by Roentgen in 1895, there had been a controversy as to their nature. They were not deflected by magnetic fields and were therefore not charged particles, but neither were they diffracted by the most closely ruled gratings, as would have been expected if they were short–wavelength radiation. In 1912 Max von Laue (1879–1950) proposed that X–rays were composed of radiation of such short wavelength that successive atomic layers in crystals might form an effective diffraction grating. The theory was tested by W. Friedrich and P. Knipping, who directed a narrow beam of X–rays at a large crystal of copper sulphate and obtained a crude diffraction pattern on a photographic plate (Figure 11.10). This crucial experiment led to the development of X–ray crystallography, which was to become such an important technique in the determination of molecular structure (Chapter 13). However, the immediate importance of X–ray diffraction was that demonstrated by W. H. and W. L. Bragg, in that it provided a method of measuring the wavelength of X–rays.

Moseley devised an apparatus which enabled the X–rays generated by firing electrons at a metallic target in a cathode ray tube to be diffracted by a crystal of potassium ferrocyanide and the resulting spectral lines to be recorded photo–graphically. Moseley found that each element produced its own characteristic set of X–ray lines, and he commented that the method 'makes the analysis of X–rays as simple as any other branch of spectroscopy'. In his first paper Moseley measured the frequencies of one of the characteristic lines in the X–ray spectra of the elements from calcium to zinc (with the exception of scandium). He found that the frequency of the lines was proportional to Q, where Q increased by a constant amount between consecutive elements when ordered according to the periodic table. Moseley continued:

'Except in the case of nickel and cobalt this is also the order of atomic weights. While, however, Q increases uniformly the atomic weights vary in an apparently arbitrary manner, so that an exception in their order does not come as a surprise. We have here a proof that there is in the atom a fundamental quantity which increases by regular steps as we pass from one element to the next. This quantity can only be the charge on the central positive nucleus, the existence of which we already have definite proof'.

Figure 11.9 Henry Gwyn Jeffreys Moseley (1887–1915)

Figure 11.10 The first X–ray diffraction photograph obtained by Friedrich and Knipping in 1912 using a crystal of copper sulphate

Moseley introduced the term *atomic number*, *N*, for the fundamental quantity to which he referred. He showed that *N* was related to *Q*, and that '*N* is the same as the number of the place occupied by the element in the periodic system'.

The following year Moseley reported on the X–ray spectra of 30 further elements lying between aluminium and gold. He confirmed that gold has an atomic number of 79, and that elements with atomic numbers 43, 61, 72 and 75 remained undiscovered. Moseley's work not only indicated how many new elements were to be expected, but also confirmed the validity of placing argon before potassium, cobalt before nickel, and tellurium before iodine, in inverse order of atomic weight. In 1915 Moseley was killed in action in the battle of Suvla Bay. The investigation he had pursued so successfully was completed in 1916 by Siegbahn and Friman, who determined the X-ray spectra of all the available metals from gold to uranium, and thereby demonstrated that the latter has an atomic number of 92.

NUCLEAR PARTICLES

After Rutherford had established that the tiny atomic nucleus contained all the positive charge and most of the mass of the atom, there was much speculation that all nuclei might be composed of positive hydrogen nuclei along with a certain number of electrons. Thus the helium nucleus would be composed of four hydrogen nuclei and two electrons, and its net positive charge of two units would be balanced by two extranuclear electrons.

In 1917 Rutherford achieved the artificial disintegration of a nucleus. His apparatus (Figure 11.11) consisted of an evacuated chamber containing the alpha emitter radium C (^{214}Bi). Alpha particles passed through a hole covered by a very thin silver plate and were detected by a zinc sulphide screen. When nitrogen gas was admitted to the chamber, the number of scintillations recorded on the screen increased. Rutherford suggested that hydrogen nuclei were being ejected from the nitrogen atoms as a result of collisions with the alpha particles. In 1920 Rutherford named the hydrogen nucleus the *proton*.

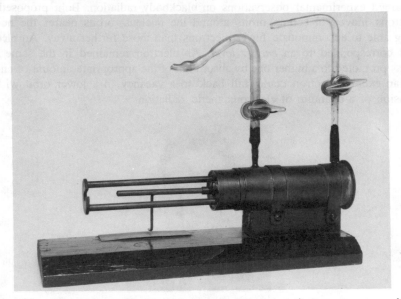

Figure 11.11 The apparatus that Rutherford used to show that nitrogen atoms were being disintegrated by alpha particles

Until 1932 it was generally accepted that the nucleus consisted of protons and electrons, but in that year the attention of James Chadwick (1891–1974) was captured by a recent experiment of Irène Joliot–Curie (1897–1956) and her husband Frédérick Joliot (1900–1958). They had found that light elements such as beryllium when bombarded with fast alpha particles from polonium emitted a very penetrating radiation. This radiation could not be deflected by a magnetic field, but when passed into paraffin wax caused protons to be ejected which had a remarkably high velocity. The Joliot–Curies were unable to provide a satisfactory explanation for the high energy and penetrating power of the radiation, but Chadwick demonstrated that it consisted of a neutral particle with approximately the same mass as the proton. Chadwick termed the new particle the *neutron*, although for a while he regarded it simply as a composite of a proton and an electron. Subsequent work showed that the neutron should be regarded as an elementary particle in its own right.

ELECTRONIC STRUCTURE

Rutherford's theory that the extranuclear electrons were circling the nucleus suffered from the objection that, according to classical theory, orbiting electrons should lose their energy and spiral inwards. The solution to the problem was suggested in 1913 by Niels Bohr (1885–1962), who introduced the concepts of the quantum theory to the motion of the extranuclear electrons. It was Max Planck (1858–1947) who in 1900 had introduced the revolutionary idea that energy is not emitted or absorbed by bodies on a continuous basis but in certain discrete units called *quanta*. The theory encountered considerable opposition, but successfully explained experimental observations on blackbody radiation. Bohr proposed that electrons moved in certain orbits around the nucleus, orbits nearer the nucleus being able to accommodate fewer electrons than those further away. An electron orbit corresponded to an *energy level*. An electron remained in the same orbit unless promoted to a higher one by absorption of the appropriate amount of energy, and an excited electron could 'fall back' to a vacancy in a lower orbit with the emission of a quantum of electromagnetic radiation.

Figure 11.12 Niels Henrik David Bohr (1885–1962)

Bohr (Figure 11.12) was born in Copenhagen, where his father was professor of physiology at the University. He obtained his doctorate at Copenhagen in 1911 with a dissertation on the electron theory of metals. While at Copenhagen he became convinced of the inability of classical electrodynamics to account for atomic phenomena. He went to Cambridge in 1911 to work with Thomson, and then moved to Manchester to collaborate with Rutherford. Bohr was one of the first to see that the chemical properties of the atom were explicable in terms of the extranuclear electrons, whereas the radioactive properties were due to the nucleus itself. Bohr arrived at the group displacement laws and the concept of isotopes before Soddy and Fajans, but Rutherford thought that the evidence did not justify the conclusions and dissuaded him from publishing. Bohr returned to Copenhagen in 1916, and two years later became the first Director of the Institute for Theoretical Physics. Under Bohr's leadership the Institute attracted many leading physicists from all over the world. After the discovery of nuclear fission, Bohr worked on the mechanics of the process and showed that the fissile isotope of uranium was ^{235}U and that ^{238}U would usually absorb slow neutrons without fission. After the outbreak of World War II he stayed in Denmark but in 1943, fearing imprisonment by the occupying Germans, he escaped to Sweden in a fishing boat. He was flown to England in the bomb bay of an unarmed plane, and was amazed to learn of the progress which had been made in the Manhattan atomic bomb project. As well as making some contributions to the project himself, Bohr was one of the first to advocate the international control of atomic weapons once the war was over. He returned to Denmark after the war and played a leading part in the foundation of CERN (European Centre for Nuclear Research).

Bohr's theory was received with a certain amount of scepticism by Rutherford, but it did have the advantage of explaining various features of atomic spectra. There had been numerous attempts to rationalise the lines observed in atomic emission spectra since the invention of the spectroscope by Bunsen and Kirchhoff in 1859 (Chapter 9). Little progress was made until 1885 when Johann Jacob Balmer (1825–1898), a Swiss school teacher, showed that the wavelengths of the four lines then known in the hydrogen spectrum could be expressed in terms of a simple equation. In 1890 Balmer's formula was rearranged by Johannes Robert Rydberg (1854–1919) to the form

$$1/\lambda = R(1/2^2 - 1/n^2)$$

Rydberg expressed the wavelengths in centimetres; thus $1/\lambda$ was the wavenumber value of the radiation, or the number of waves per centimetre. R, the Rydberg constant, had the value 109 720 cm^{-1}. The wavelengths of the four lines corresponded to values of n ranging from 3 to 6. Until 1908 the hydrogen spectrum had been found to contain only this series of lines in the visible region (the *Balmer series*), but over the next two decades a series was found in the ultraviolet by Lyman, and three more were found in the infrared by Paschen, Brackett and Pfund. The wavenumber values of the lines in these new series corresponded with the substitution of the integer values of 1, 3, 4 and 5, respectively, in place of 2 in the Rydberg formula.

Bohr assumed that the angular momentum of the electron in an orbit had to be an integral multiple of $h/2\pi$, where h is Planck's constant. He was therefore able to derive an equation for the energy evolved when an electron in a hydrogen atom changes its orbit. The equation was of the same form as the Rydberg equation and contained a constant term made up of known quantities such as Planck's constant and the mass and charge of the electron. When Bohr used the accepted values for these quantities to evaluate the constant in his equation, he obtained a value remarkably close to the Rydberg constant. By making appropriate modifications to his equation, Bohr was able to predict the frequencies of the lines in the spectra of other one–electron species such as He^+, Li^{2+}, Be^{3+}, etc. These spectra were later observed in electric discharges and the frequencies were found to be very close to Bohr's predictions.

Bohr also attempted to write electronic structures (in terms of orbits) for elements up to chromium. However, the theoretical treatment of multi–electron systems proved to be very complex, and Bohr assigned electronic structures for elements principally on the basis of chemical properties. Thus lithium was assigned a structure of (2,1) but nitrogen was presumed to be (4,3) because of its trivalency. Over the next few years, Bohr, along with Arnold Sommerfeld (1868–1951), William Wilson and Charles Bury, refined electronic structures on the basis of both spectral and chemical behaviour.

The observation of spectral lines under high resolution revealed that many possessed a fine structure, and this led to the concept of electronic *sublevels*. These were named s, p, d and f levels, the letters having their origin in the atomic spectra of the alkali metals in which four series of lines were observed, which were known as *sharp, principal, diffuse* and *fundamental*. In 1896 some lines had been found to be split in a magnetic field by Pieter Zeeman (1865–1943), and this phenomenon was now explained in terms of electron spin. Each electron was now described in terms of four quantum numbers: principal (n), orbital (l), magnetic (m) and spin (s). In 1925 Wolfgang Pauli (1900–1958) put forward his *exclusion principle*, which stated that no two electrons in a given atom could have all four quantum numbers the same.

QUANTUM MECHANICS

In 1924 Louis de Broglie (1892–1987) introduced ideas which were to revolutionise our concept of the electron and other fundamental particles. The quantum theory had proposed that electromagnetic radiation was particulate in some of its properties, but only a wave theory could explain diffraction. De Broglie proposed that both the wave and particle theories should be accepted. He also suggested that matter should be regarded as not only particulate in nature, but should also be considered to have a wave nature as well. The validity of this viewpoint was demonstrated in 1927 with Davisson and Germer's experimental observation of the diffraction of electrons by crystals.

Shortly after de Broglie's theory appeared, new mathematical treatments, known collectively as *quantum mechanics*, were introduced to describe the behaviour of electrons in atoms. Erwin Schrödinger (1887–1961) developed *wave mechanics*, and Werner Heisenberg (1901–1976) used a different approach called *matrix mechanics*. Heisenberg's treatment led to the *uncertainty principle*, which stated that it is impossible simultaneously to determine precisely both the position and the velocity of the electron. Other workers in the field were Max Born (1882–1970), Pascual Jordan and Paul Dirac (1902–1984). Although the approaches of Schrödinger and Heisenberg seemed very different, it was ultimately realised that there is a fundamental unity between them.

The Schrödinger wave equation is one of the cornerstones of modern theoretical chemistry. From a consideration of a standing wave in three dimensions, Schrödinger derived an expression involving the wavefunction Ψ, the coordinates x, y and z and the total and potential energies E and V (Figure 11.13). Schrödinger suggested that the square of the wavefunction (Ψ^2) represented electron density at a particular point, but Born suggested that it represents the *probability* of finding the electron at that point. Solution of the wave equation for 90 per cent probability yields an envelope within which there is that chance of finding the electron. Such envelopes are known as *orbitals*. Only approximate solutions to the wave equation are possible, but from these have come our models of the various types of orbital. Each type of sublevel (s, p, d, f) consists of a characteristic number of orbitals (one, three, five, seven, respectively), and each orbital can hold two electrons with opposite spins.

$$\frac{\delta^2 \psi}{\delta x^2} + \frac{\delta^2 \psi}{\delta y^2} + \frac{\delta^2 \psi}{\delta z^2} + \frac{8\pi^2 m}{h}(E - V)\psi = 0$$

Figure 11.13 The Schrödinger wave equation

THE CHEMICAL BOND

The first attempt to explain the chemical bond in terms of electrons was made in 1904 by Thomson in his paper on the structure of the atom. He proposed that transfer of corpuscles (electrons) occurred from one atom to another at the moment of compound formation. He continued:

'The electronegative atoms will thus get a charge of negative electricity and the electropositive atoms one of positive, the oppositely charged atoms will attract each other, and a chemical compound of the electropositive and electronegative atoms will be formed'.

The application of Thomson's ideas to less polar compounds presented serious difficulties. There was no escape from the conclusion that certain atoms, e.g. carbon, could either accept or donate electrons depending on the nature of the atoms it was joined to. Thus, in methane, the carbon atom was supposed to receive electrons from the hydrogen atoms; but, in tetrachloromethane, it was carbon which was the provider of electrons. This reasoning led to the conclusion that for certain compounds *electronic isomers* (or *electromers*) might exist (Figure 11.14), but none could be isolated.

$$
\begin{array}{cc}
\ominus & \oplus \text{ Cl} \\
\text{N} \ominus & \oplus \text{ Cl} \\
\ominus & \oplus \text{ Cl}
\end{array}
\qquad\qquad
\begin{array}{cc}
\oplus & \ominus \text{Cl} \\
\text{N} \oplus & \ominus \text{Cl} \\
\oplus & \ominus \text{Cl}
\end{array}
$$

Figure 11.14 Proposed electromers of nitrogen trichloride

In 1916, after Moseley had provided evidence of the number of extranuclear electrons in an atom and Bohr had introduced the idea of electrons being arranged in shells, Gilbert Newton Lewis (1875–1946) published an important paper entitled *The Atom and the Molecule*. Lewis was developing some ideas which he had jotted down as early as 1902 (Figure 11.15). He had imagined the electrons to be arranged in concentric cubes, and his sketch implies that he considered that bonding occurred between atoms as a result of the transfer of sufficient electrons to complete a cube. Lewis discussed his ideas with colleagues and students but published nothing on this subject before his paper of 1916.

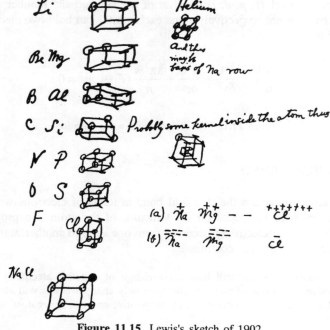

Figure 11.15 Lewis's sketch of 1902

In his 1916 paper Lewis stated clearly that bonding in polar compounds occurred as a result of electron transfer, which yielded oppositely charged ions with complete cubes of electrons. In the same year, Walther Kossel (1888–1956), in

Germany, also realised the importance of eight electrons but imagined the shells to be circles rather than cubes. His picture of bonding in polar compounds was essentially the same as that of Lewis.

Lewis not only considered very polar compounds such as sodium chloride, but also extended his discussion to entities such as hydrogen chloride and iodine. Lewis proposed that in such cases electrons could be shared in pairs between the atoms by the two cubes sharing a side or a face. In the same paper Lewis introduced the familiar symbolism of dots to represent electrons in the outer shells (Figure 11.17). He also suggested that polar molecules resulted from unequal sharing of electron pairs, and that bonding in compounds such as sodium chloride could be regarded as an extreme case of unequal sharing.

Lewis (Figure 11.16) was born in Weymouth, Massachusetts. After gaining his Ph.D. from Harvard in 1899, he travelled to Europe, spending time in Leipzig and Göttingen. Among the great European chemists he worked with were Ostwald and Nernst. After returning to the United States, he was an instructor at Harvard for a brief period, before going to Manila in the Philippines to work for the Bureau of Science. He returned to the United States in 1905 and worked at the Massachusetts Institute of Technology until 1912. In that year he became chairman of the chemistry department at the University of California at Berkeley, where he remained until his death. Lewis is remembered not only for his work on bonding but also for his contributions to thermodynamics. His careful experiments provided experimental justification for the third law of thermodynamics, which states that entropy becomes zero at absolute zero for all perfect crystals. He also widened the definitions of the terms *acid* and *base* (Chapter 13).

Figure 11.16 Gilbert Newton Lewis (1875–1946)

Lewis's ideas were extended in 1919 by Irving Langmuir (1881–1957), who drew a clearer distinction between the bonding in the two types of compound and introduced the terms *electrovalency*, *covalency* and *octet*. Like Lewis, he also represented covalent compounds by means of joined cubes (Figure 11.18). Other developments soon followed. In 1920 Wendell M. Latimer and Worth H. Rodebush introduced the concept of the *hydrogen bond* to explain the association between molecules in liquids such as water and acetic acid. The *coordinate bond*, in which both shared electrons originate from the same atom, was proposed in 1927 by Neil Vincent Sidgwick (1873–1952) to explain the bonding between the central metal ion and the ammonia molecules in ammine complexes such as $[Cr(NH_3)_6]^{3+}$. This was an extension of another of Lewis's ideas. In his paper of 1916 Lewis had explained the bonding in the ammonium ion as a sharing of the lone pair of electrons of an ammonia molecule with a hydrogen ion.

$$H:H \qquad H:\overset{..}{\underset{..}{O}}:H \qquad H:\overset{..}{\underset{..}{I}}: \qquad :\overset{..}{\underset{..}{I}}:\overset{..}{\underset{..}{I}}:$$

Figure 11.17 Lewis's 'dot' formulae

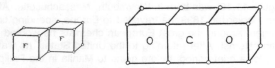

Figure 11.18 Langmuir's diagrams of joined cubes

In the 1920s two British chemists, Robert Robinson (1893–1975) and Christopher Ingold (1893–1970), separately started to develop explanations of the course of organic reactions based upon the Lewis–Langmuir concepts of bonding in covalent compounds. These early ideas on reaction mechanisms produced new insights into the bonding in the organic molecules themselves. In 1922 Robinson introduced the idea of the concerted movement of pairs of electrons in conjugated systems, and showed how one Kekulé formula for benzene could be converted into the other by the movement of electron pairs.

In 1924 F. G. Arndt drew attention to the fact that the reactions of the γ–pyrones were not consistent with their classical formula I (Figure 11.19). Utilising the recently introduced idea of the *zwitterion*, he proposed an alternative structure (II). Arndt's vital contribution was to suggest that neither formula provided an accurate representation of the molecule, but that the true structure was some kind of intermediate state.

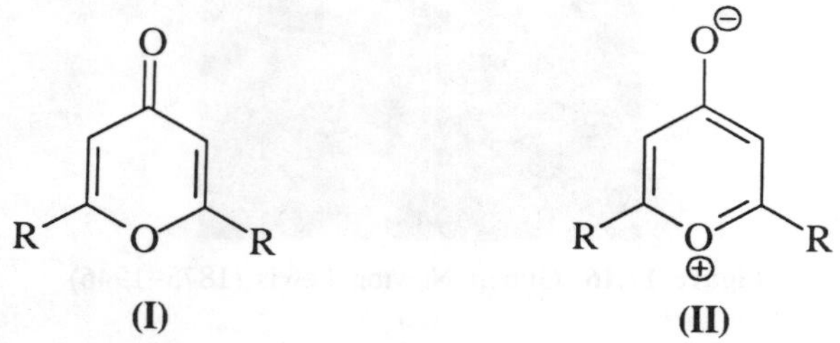

(I)   (II)

Figure 11.19 Two possible formulae for the γ–pyrones

Arndt's concept was elaborated in 1926 by Ingold, who pointed out that dimethylaniline might not only have structure III (Figure 11.20) but that there might also be a contribution from structure IV. Ingold further proposed that, because of the electron–attracting power of nitrogen, the nitrogen atom in structure III should act as the negative end of the dipole, but that the polarity is reversed in structure IV. Ingold noted that dipole moment measurements should be capable of establishing whether structure IV makes a contribution. When the measurements were made, it was clear that the structure of dimethylaniline is indeed intermediate between forms III and IV. Ingold proposed the term *mesomerism* to describe the intermediate character of the bonding. Until this time many organic chemists had doubted if physical chemistry was capable of yielding any results of value for their discipline, but now *physical organic chemistry* rapidly began to make important contributions, especially in the areas of molecular structure and reaction mechanisms.

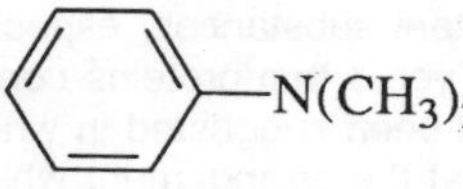

Figure 11.20 Two possible formulae for dimethylaniline

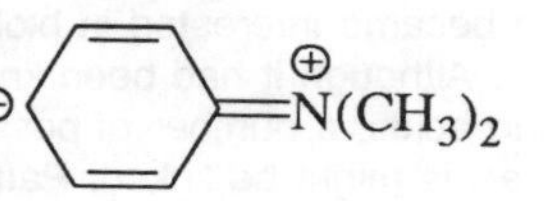

VALENCE BOND AND MOLECULAR ORBITAL THEORIES

The new science of quantum mechanics proved to be capable of providing an explanation of the concept of mesomerism as well as the more familiar aspects of bonding such as directed valency. The two approaches which have been developed are known as the *valence bond theory* and the *molecular orbital theory*.

The valence bond approach uses the wavefunctions of the valence electrons of two approaching atoms to calculate the potential energy of the system as a function of the interatomic distance (Figure 11.21). This method was first applied to the hydrogen molecule by Walter Heitler (*b.* 1904) and Fritz London (1900–1954) in 1927, and subsequent refinements to the calculations have produced remarkably good values for the bond energy and the bond length.

More than anyone else it has been Linus Pauling (*b.* 1901) who has been responsible for the development and application of the valence bond theory. In the early 1930s he deduced from quantum mechanics the tetrahedrally directed valencies of carbon, and he introduced the concept of the hybridisation of atomic orbitals. He introduced the idea of *resonance* as the quantum–mechanical counterpart of mesomerism. The wavefunction for the molecule must contain terms for all possible structures, and the molecule is said to *resonate* between them. In 1933 Pauling described the benzene molecule as a resonance hybrid between the two Kekulé structures and the three possible Dewar structures (Figure 11.22).

Pauling and John C. Slater applied the valence bond theory to coordination compounds, describing the bonding in terms of overlap of an orbital of the ligand containing a lone pair of electrons with a vacant hybridised orbital on the central metal ion. This model gave a satisfactory explanation of both the geometry and magnetic properties of many complexes, but was unable to explain their spectral behaviour.

Pauling (Figure 11.23) was born in Oregon, and was a student at the Oregon Agricultural College. His father had died when Pauling was nine, and he took a series of jobs to support himself while he studied. He so distinguished himself at chemistry that he was given some teaching duties at the College while still an undergraduate. Pauling then moved to the newly opened California Institute of Technology to study for his Ph.D. He worked on X–ray diffraction, solving the structure of many minerals. In 1925 Pauling went to Europe, where he spent two years with Arnold Sommerfeld, who was working on the new quantum mechanics. On his return to the United States, Pauling performed his famous work on bonding, which resulted in 1939 in the publication of his book *The Nature of the Chemical Bond*. In the 1930s Pauling became interested in biologically important substances, especially proteins. Although it had been known for many years that proteins consisted of amino acids, a number of possible ways had been suggested in which the amino acids might be linked. Pauling argued that the arrangement which best fitted the data available at the time was that the amino acids were joined together in long chains. He studied amino acids and small peptides by X–ray diffraction, and in 1950 he used the data he had acquired to propose the alpha–helical structure for the peptide chain. In the 1940s Pauling studied samples of haemoglobin isolated from normal humans and those suffering from sickle–cell anaemia. He showed that the two haemoglobins moved differently in electrophoresis and therefore had different molecular structures. This was the first example of a *molecular disease*. Pauling is one of the select few who have won two Nobel prizes. He was awarded the Chemistry Prize in 1954 and the Peace Prize in 1962 for his campaigns against the proliferation and testing of nuclear weapons. His views that large doses of vitamin C are effective in combatting the common cold and cancer have been criticised by the medical fraternity. He has remained active in his eighties, and has published papers on high–temperature superconductivity. His work has been characterised throughout by brilliant intuition and insight, and he is undoubtedly one of the greatest chemists of the twentieth century.

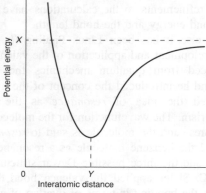

Figure 11.21 Plot of potential energy against interatomic distance for a diatomic molecule

Robert Mulliken (1896–1986) was the chief originator of the molecular orbital approach. This theory rests on the postulate that, when a compound is formed, electrons move in new orbitals which extend over several atomic nuclei. The mathematical approximation used is known as the LCAO (linear combination of atomic orbitals) method. In the molecular orbital theory, a molecule is considered to possess a series of orbitals of increasing energy. Electrons occupy the lower energy *bonding orbitals*, while there will be vacancies in the higher energy *antibonding orbitals*. Just as atomic spectra can be explained in terms of the movement of electrons between atomic orbitals, the absorption spectra of molecules in the ultraviolet and visible regions are best explained in terms of electron movement between molecular orbitals.

Figure 11.22 The five structures contributing to the normal state of the benzene molecule

Figure 11.23 Linus Carl Pauling (*b.* 1901)

It was Erich Hückel (1896–1980) who first attempted to apply the molecular orbital theory to organic molecules. He introduced the terms *sigma* (σ) and *pi* (π)

bonding. He applied the molecular orbital theory to aromatic molecules in 1931 when he derived the rule that for a cyclic compound to display aromatic character, it must possess $(4n + 2)$ π electrons, where *n* is an integer.

The spectral behaviour of coordination complexes was first satisfactorily explained in terms of the *crystal field theory*, developed in 1935 by John H. van Vleck from the work of Hans Bethe. On this theory the bonding is regarded as being essentially electrostatic in nature, there being an attraction between the central positive ion and the negative or partially negative ligand. Later, molecular orbital concepts were introduced into the crystal field theory to produce the modern *ligand field theory*.

Both the valence bond and molecular orbital theories continue to be used today, and both involve approximations. Neither approach can claim to be final, and the chemist uses the method most suited to the problem in hand.

12

Inorganic Chemistry

With the rapid development of carbon chemistry after 1860, a considerable number of chemists, especially in Germany, devoted all their attention to this branch of the subject. Chemistry therefore became divided into organic and inorganic specialisms, and towards the end of the century physical chemistry began to emerge as a third discipline. The progress in carbon chemistry outshone that made in other areas until the 1890s, when new discoveries and theories relating to coordination compounds signalled the coming of age of inorganic chemistry.

NEW ELEMENTS

We have seen how the number of known elements continued to grow during the second half of the nineteenth century. Powerful tools in the detection of new elements were the spectroscope, invented in 1860 (Chapter 9), and radioactivity, first exploited by Marie and Pierre Curie in 1898 (Chapter 11). Mendeleev's periodic classification had predicted the existence of gallium, scandium and germanium, but the discovery of the rare gases had come as a surprise (Chapter 9). In contrast, the existence of fluorine had been suspected for many years, but its extreme reactivity resulted in its isolation being delayed until 1886.

The mineral fluorspar had been used as a flux since the middle ages (the name comes from the Latin *fluor* meaning *flow*), and in the *Traité* Lavoisier had included the fluoric radical as a simple substance (element) as yet unknown. Scheele obtained hydrofluoric acid in 1771. In 1813 Ampère pointed out the many similarities between the fluoric and muriatic (chlorine) compounds. He suggested the unknown element be called fluorine, and in the same year Humphry Davy tried to prepare it by the electrolysis of hydrofluoric acid. The corrosive nature of the acid presented insuperable problems, and it was to be another 73 years before the isolation of the element was achieved. During that time the toxic nature of fluorine compounds was to be responsible for the deaths of at least two chemists and was to ruin the health of many more. Success was finally achieved by Henri Moissan (1852–1907), who electrolysed potassium fluoride dissolved in anhydrous hydrofluoric acid using platinum apparatus.

The rare earth elements (lanthanum and the ensuing fourteen elements) proved very troublesome to isolate. The difficulties stemmed partly from the fact that these elements are so similar in properties (and occur together) that their separation was extraordinarily difficult. The metals are also fairly reactive, and most were known for many years as the oxide (or *earth*) before they were successfully reduced to the metal. The term *rare earth element* is something of a misnomer, as the least abundant of the group, thulium, is as abundant as bismuth in the earth's crust and more common than arsenic, cadmium and mercury.

The first discoveries relating to the rare earths were made at the end of the eighteenth century at the time when careful analytical work on minerals was revealing the presence of new elements. In 1794 Johan Gadolin (1760–1852) was investigating a mineral from Ytterby, near Stockholm, and obtained an earth which was subsequently named *yttria*. In 1803 Klaproth isolated an earth somewhat similar to yttria from another Swedish mineral, and about the same time Berzelius and Hisinger also obtained this new earth and called it *ceria*.

Both yttria and ceria proved to be complex mixtures of oxides, and the separation of individual compounds from them proceeded in fits and starts throughout the nineteenth century. By 1907 yttria and ceria had yielded the oxides of yttrium, scandium, lanthanum and thirteen of the following fourteen elements.

Although the periodic classification had suggested the existence of gallium, scandium and germanium, it was no help in deciding how many rare earth elements should be expected. Moseley's technique of X-ray spectroscopy (Chapter 11), discovered in 1913, was able to pinpoint the remaining gaps in the table with great precision. The X-ray spectral evidence showed that element number 61 was missing from the rare earth series, and numbers 43, 72 and 75 were missing from the main body of transition elements. Elements 43 and 61 (technetium and promethium) have no naturally occurring isotopes and have been produced artificially. Elements 72 and 75 (hafnium and rhenium) were isolated in 1923 and 1925 respectively. Bohr's periodic table of 1920 left gaps for the above four elements, and also those of atomic numbers 85 and 87. Both these occur only as short-lived radioisotopes. Element 87 (francium) was identified in 1939 by Marguerite Perey (1909–1975) among the decay products of actinium–227, while element 85 (astatine) was produced artificially in 1940.

All the artificial elements are, of course, radioactive, and the first to be prepared was technetium. It was produced in vanishingly small quantity in 1939 by Segrè and Perrier by bombarding element 42 (molybdenum) with deuterons for several months. It was also detected the following year among uranium fission products. A few years later, the same source yielded promethium, the missing rare earth element.

The elements with atomic numbers higher than 92, the *transuranium elements*, are generally produced by bombarding suitable nuclei either with neutrons in a nuclear reactor or with accelerated positive particles. Elements 99 and 100 (einsteinium and fermium) were initially detected among the products of the first hydrogen bomb explosion in 1952. All the transuranium elements are artificial, with the exception of plutonium, one isotope of which is found in natural uranium and is formed as a result of neutron bombardment in the mineral deposit. Using bombardment techniques artificial radioisotopes of all the elements in the periodic table have now been prepared, and some have found application in industry and medicine.

VARIABLE VALENCY

We have seen how, after 1860, organic chemists were increasingly successful in their attempts to assign molecular structures to organic compounds. Although Frankland and Couper had considered variable combining power to be a possibility, Kekulé insisted that the atomicity (i.e. valency) of an atom was as constant and unchangeable as its atomic weight. While this notion was tenable in the field of organic chemistry, it met with serious difficulties in inorganic chemistry, and considerable ingenuity was exercised in devising formulae for the compounds of the increasing number of elements which appeared to display variable valency. Dimeric formulae were sometimes used; for example, in 1870 it was proposed by Thorpe that the newly discovered $VOCl_2$ existed as a double molecule in order to preserve the supposed trivalency of vanadium (Figure 12.1). Another device to prevent atoms having unacceptably high valencies was the incorporation of chains of atoms into formulae. This is exemplified by Kekulé's suggested structure for sulphuric acid, which accorded sulphur a valency of two (Figure 12.2).

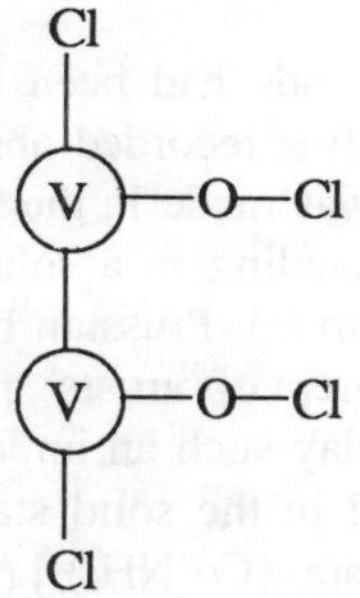

Figure 12.1 Thorpe's structure for $VOCl_2$

$$H-O-O-S-O-O-H$$

Figure 12.2 Kekule's structure for H_2SO_4

The dimerisation and chain theories were clearly powerless to provide an explanation in cases where an atom apparently displaying an abnormally high valency is bonded solely to monovalent atoms (e.g. PCl_5, NH_4Cl). Such cases were explained by the ever resourceful Kekulé in 1864 in terms of his *molecular compound theory*, which stated that two molecules might adhere by means of relatively weak forces. Thus phosphorus pentachloride was formulated $PCl_3.Cl_2$, and ammonium chloride $NH_3.HCl$. Evidence for Kekulé's views was provided by the fact that both phosphorus pentachloride and ammonium chloride were found to dissociate in the vapour phase into the constituent molecules indicated in the formula.

All these attempts to resist variable valency came under attack. In principle, it is possible to establish whether or not a compound is dimeric by determining its molecular weight. The only method available in the 1860s and 1870s was that of vapour density, but many inorganic compounds were found to be too involatile for measurements to be made. In cases where determinations were possible, the results were not always consistent with the dimeric formulae which had been proposed. Just as organic structures were proposed or verified using arguments based on synthesis, similar reasoning applied to inorganic compounds often appeared to be incompatible with dimeric or chain formulae. The molecular compound theory also encountered difficulties. In 1876 Thorpe prepared phosphorus pentafluoride, PF_5, and found that it showed no evidence of dissociation in the vapour phase.

Kekulé found himself increasingly isolated in his advocacy of constant valency, and by the time of his death in 1896 he had few supporters. Although the concept of constant valency was in retreat for reasons already noted, the great revolution in thought in inorganic chemistry occurred as a result of a study of a relatively new class of substances, known today as *coordination compounds*.

COORDINATION COMPOUNDS

A few coordination compounds had been known for many years as a result of accidental discoveries. The first recorded observation of the deep blue colour of the tetraamminecopper(II) ion was made in the sixteenth century by Andreas Libavius when he left some brass standing in a solution of lime water saturated with sal ammoniac (ammonium chloride). Prussian blue was obtained accidentally in 1704 by Diesbach, a manufacturer of artists' pigments. It was the cobalt–ammine complexes which were to play such an important role in chemical theory, and the first of these to be isolated in the solid state was the compound now known as hexaamminecobalt(III) oxalate, $[Co(NH_3)_6]_2(C_2O_4)_3$. This was described in 1822 by Leopold Gmelin (1788–1853).

Serious study of the cobalt–ammine complexes was initiated by Frederick Augustus Genth (1820–1893) (Figure 12.3) as a result of an accidental discovery. In 1847 Genth was Bunsen's assistant at the University of Marburg, and, when demonstrating the procedure for the qualitative analysis of metals in solution, he found that there was no potassium hydroxide available to add to the mixture after the metals in analytical group II had been precipitated. He therefore used ammonia instead, and left the solution to stand until the next class, which was after a vacation. Upon his return, Genth noticed that the solution had deposited beautiful coloured crystals, which he found contained cobalt.

Genth emigrated to the United States before he was able to complete his investigations, and the first account of his work was published in his adopted country in 1851. He described salts of the types now designated by the formulae $[Co(NH_3)_6]X_3$ and $[Co(NH_3)_5H_2O]X_3$. The following year the American Oliver

Wolcott Gibbs (1822–1908) started investigating coordination compounds, and in 1856 Genth and Gibbs published a lengthy paper in which they described a total of 35 cobalt–ammine complexes. These included those discussed in Genth's paper of 1851 as well as salts of the type $[Co(NH_3)_5Cl]X_2$ and $[Co(NH_3)_5NO_2]X_2$. Gibbs and Genth concluded their paper with the words 'we invite the attention of chemists to a class of salts which for beauty of form and color, and for abstract theoretical interest, are almost unequalled either among organic or inorganic compounds'.

Figure 12.3 Frederick Augustus Genth (1820–1893)

Even before the work of Gibbs and Genth, chemists had produced theories to rationalise the constitution of the relatively small number of coordination compounds then known. In 1837 Thomas Graham (1805–1869) suggested that the ammines were ammonium salts in which some of the hydrogen was replaced by a metal while in 1856 Carl Ernst Claus (1796–1864) proposed that the ammines were conventional metal salts which contained ammonia 'passively' bound to them. Somewhat later Kekulé denoted the ammines as molecular compounds. Thus hexaamminecobalt(III) chloride became $CoCl_3.6NH_3$. Such a formulation was clearly unsatisfactory, as no ammonia is evolved on gentle heating, no ammonium sulphate is formed with sulphuric acid, and the addition of base fails to precipitate cobalt(III) hydroxide.

THE BLOMSTRAND–JØRGENSEN CHAIN THEORY

The most successful theory of coordination compounds prior to that of Alfred Werner was advanced in 1869 by Christian Wilhelm Blomstrand (1826–1897) (Figure 12.4). In order to account for the lack of reactivity of ammonia in the ammines, he proposed that chains of ammonia molecules linked the metal atom to other parts of the molecule (Figure 12.5). Blomstrand thereby assigned nitrogen a valency of five, despite the disapproval of Kekulé. The chains of ammonia molecules are reminiscent of some of the chain structures discussed earlier, and were advanced at a time when chains of CH_2 groups were becoming commonplace in organic formulae.

Figure 12.4   Christian Wilhelm Blomstrand (1826–1897)

$$Co_2\begin{cases} NH_3\!-\!NH_3\!-\!Cl \\ NH_3\!-\!NH_3\!-\!Cl \\ NH_3\!-\!NH_3\!-\!Cl \\ NH_3\!-\!NH_3\!-\!Cl \\ NH_3\!-\!NH_3\!-\!Cl \\ NH_3\!-\!NH_3\!-\!Cl \end{cases}$$

Figure 12.5 Blomstrand's formula for hexaamminecobalt(III) chloride. At this time, double formulae were used for the cobalt–ammines since cobalt(III) chloride was assumed to be Co_2Cl_6 by analogy with iron(III) chloride, which had been shown to be Fe_2Cl_6 from vapour density measurements

The greatest advocate of the chain theory was the Dane, Sophus Mads Jørgensen (1837–1914), professor of chemistry at the University of Copenhagen (Figure 12.6). Jørgensen spent almost 30 years investigating the coordination compounds of cobalt, chromium, rhodium and platinum. By 1890, it was possible to estimate the molecular weight of an involatile solute by the depression it produced in the freezing point when dissolved in a solvent. Using this method, Jørgensen showed that the dimeric formulae of the cobalt–ammine salts should be replaced by monomeric ones (Figure 12.7). He introduced several refinements to Blomstrand's theory to accommodate the many new compounds which he had prepared. All the time he was building up a mass of evidence which Werner was to use to support his revolutionary new theory.

Figure 12.6 Sophus Mads Jørgensen (1837–1914)

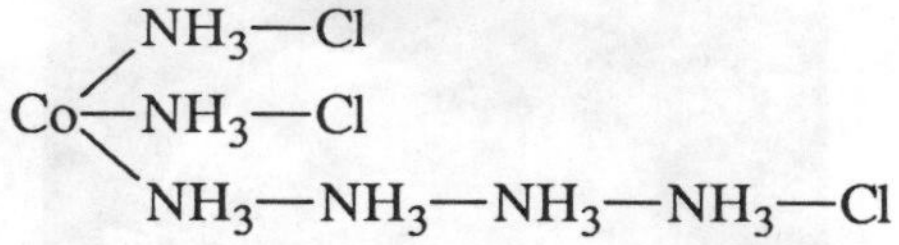

$$
\text{Co}
\begin{cases}
\text{NH}_3\text{—Cl} \\
\text{NH}_3\text{—Cl} \\
\text{NH}_3\text{—NH}_3\text{—NH}_3\text{—NH}_3\text{—Cl}
\end{cases}
$$

Figure 12.7 Jørgensen's formula for hexaamminecobalt(III) chloride

WERNER'S COORDINATION THEORY

The theory which was destined to replace the Blomstrand–Jørgensen chain theory and to form the basis for all subsequent consideration of coordination compounds was first announced in 1893 by Alfred Werner (1866–1919). Werner was no respecter of authority, and, while in the early 1890s Kekulé was widely regarded as one of the greatest living chemists, Werner had no hesitation in challenging Kekulé's views. Some of Werner's earlier work was in the field of organic stereochemistry and he questioned Kekulé's concept of rigidly directed valencies. In 1892 he delivered a lecture reviewing the various theories concerning the structure of benzene, and here again he was critical of Kekulé's view of valence forces. It was not surprising that he was unimpressed by Kekulé's molecular compound theory, and his attention was drawn to the current theories of the metal–ammonia salts while preparing another lecture in 1892. He continued to consider the problem and he claimed that, as with Kekulé, the solution came to him in the form of a dream.

Werner (Figure 12.8) was born in the French town of Mulhouse, but when he was five years old he acquired German citizenship when Mulhouse was incorporated into Germany as a result of the Franco–Prussian war. After his military service, Werner studied at the Federal Institute of Technology in Zurich. There he obtained his first and doctor's degrees, and after a further three years of research, during which his famous theory was conceived, he moved to the University of Zurich to teach organic chemistry. He remained at the University for the rest of his career, and he became a Swiss national soon after his marriage to a Swiss woman in 1894. Over the years he became more and more absorbed in his work, and this was accompanied by an increasing dependence on alcohol. He died at the early age of 52. It is hard to think of another figure in the recent history of chemistry who made a particular area of the subject more completely his own than Alfred Werner. Not only did he provide the foundation on which all subsequent work in coordination chemistry would be based, but he also demonstrated that stereoisomerism is a general phenomenon and not restricted to carbon-containing molecules. Under Werner's leadership inorganic chemistry acquired something of the dynamism and excitement which had for many years been associated with organic chemistry.

Figure 12.8 Alfred Werner (1866–1919)

Werner proposed that metals displayed two types of valency. One he called the *Hauptvalenz* or primary valency, which is *ionogenic*, resulting in the production of ions in solution. The other he called the *Nebenvalenz* or secondary valency, and this is *non–ionogenic*. Each metal atom with a particular primary valency has associated with it a definite number of secondary valencies which must be satisfied. This number Werner called the *coordination number*. The secondary valencies of the metal can be satisfied either by anions or by neutral molecules such as ammonia. These groups coordinated to the metal ion are now called *ligands*. The resulting entity is called a *complex*, which usually exists as a discrete ion in solution. The most common coordination numbers are 6 and 4, and Werner suggested that the configuration of the complex is octahedral in the first case and either square planar or tetrahedral in the second. In terms of this theory, hexaamminecobalt(III) chloride is represented as $[Co(NH_3)_6]Cl_3$.

Werner had performed no research on coordination chemistry, and his theory rested on experimental data which had been gathered by others, especially Jørgensen. Werner devoted the rest of his life to the justification of his bold propositions. Not surprisingly, Jørgensen was opposed to the new theory, and during the resulting controversy, conducted without bitterness on either side, many new discoveries were made.

Werner sought to provide evidence for his theory chiefly through studies on conductivity and isomerism. Friedrich Kohlrausch had shown that molecular conductivity (μ) provides a method of determining the number of ions in a soluble compound. In 1893 Werner and Arturo Miolati started to measure the molecular conductivities of a number of cobalt complexes. Some of their results, along with the rival Werner and Jørgensen formulae for the compounds, are listed in Table 12.1.

Table 12.1 The proposed formulae of four cobalt–ammine complexes

Formulae proposed		Experimental data	
Werner	Jørgensen	Number of ions	Fractions of chloride precipitated with $AgNO_3$
$[Co(NH_3)_6]Cl_3$	$Co{<}^{NH_3-Cl}_{NH_3-Cl}$ $NH_3-NH_3-NH_3-NH_3-Cl$	4	1
$[Co(NH_3)_5Cl]Cl_2$	$Co{<}^{Cl}_{NH_3-Cl}$ $NH_3-NH_3-NH_3-NH_3-Cl$	3	⅔
$[Co(NH_3)_4Cl_2]Cl$	$Co{<}^{Cl}_{Cl}$ $NH_3-NH_3-NH_3-NH_3-Cl$	2	⅓
$[Co(NH_3)_3Cl_3]$	$Co{<}^{Cl}_{Cl}$ $NH_3-NH_3-NH_3-Cl$	0	0

In Jørgensen's version of the chain theory, chlorine was capable of ionising and forming an immediate precipitate with silver nitrate when joined to ammonia but not when joined directly to the metal. Werner maintained that each of the ionisable chlorines was bound by a primary ionogenic valency. Both the Werner and the Jørgensen formulae are consistent with the experimental data for the first three compounds in Table 12.1. However, the fourth compound was found to be a non-electrolyte and none of the chlorine could be precipitated, results consistent with Werner's formula but not with Jørgensen's. Werner also prepared a series of complexes in which the ammonia molecules in the hexaamminecobalt(III) ion, $[Co(NH_3)_6]^{3+}$, were progressively replaced by the nitrite ion. The conductivity fell to almost zero in $[Co(NH_3)_3(NO_2)_3]$ and then rose again in $[Co(NH_3)_2(NO_2)_4]^-$.

In 1889 Jørgensen had prepared two isomers with the formula CoCl$_3$.2en, where en is ethylenediamine, $H_2N.CH_2.CH_2.NH_2$. Jørgensen believed that these were structural isomers, but Werner maintained that they were two geometric isomers of the octahedral complex ion (Figure 12.9). In fact Werner cited the existence of two compounds with this formula as evidence for the octahedral configuration in complexes with a coordination number of 6.

Jørgensen replied that, if Werner's theory was correct, geometric isomerism should also be observed in compounds such as CoCl$_3$.4NH$_3$, which Werner would denote as $[Co(Cl)_2(NH_3)_4]Cl$. Gibbs and Genth had reported the preparation of one compound with this formula in 1857 (actually the *trans* isomer), but all attempts to isolate a second isomer were unsuccessful until 1907, when Werner prepared the unstable *cis* isomer (Figure 12.10). At this point Jørgensen promptly acknowledged the validity of Werner's theory. Over the years Werner and his students prepared 53 series of cobalt and chromium complexes displaying geometric isomerism.

A consequence of the theory of octahedral coordination is that certain ions should exist as a pair of optical isomers. The first ion to be resolved was that of $cis-[Co(Cl)(NH_3)(en)_2]^{2+}$ (Figure 12.11) in 1911 by Werner and his American student Victor King (1886–1958). The method used was to form a pair of diastereoisomeric salts with the optically active bromocamphorsulphonate ion, which were then separated by fractional crystallisation.

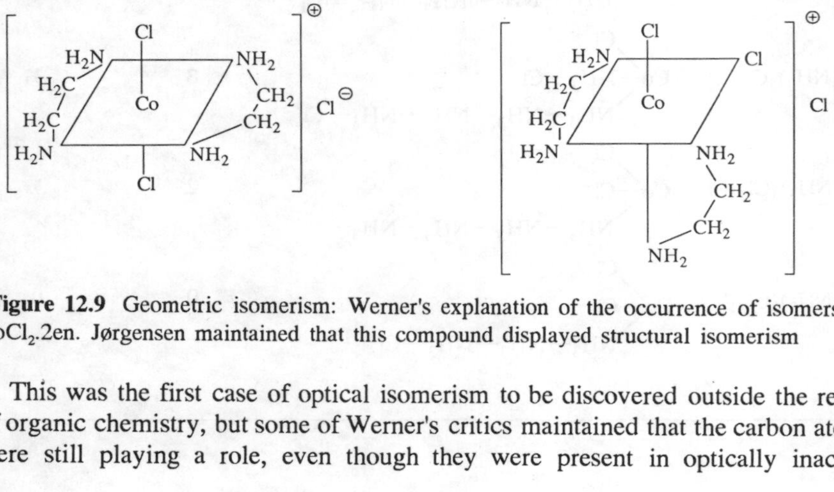

Figure 12.9 Geometric isomerism: Werner's explanation of the occurrence of isomers of CoCl$_2$.2en. Jørgensen maintained that this compound displayed structural isomerism

This was the first case of optical isomerism to be discovered outside the realm of organic chemistry, but some of Werner's critics maintained that the carbon atoms were still playing a role, even though they were present in optically inactive ethylenediamine molecules. Those who held this view were finally confounded when in 1914 Werner resolved a complex containing the cobalt(III) ion joined to three carbon–free bidentate ligands (Figure 12.12). Werner and his students finally succeeded in resolving more than forty series of optically active cationic and anionic complexes.

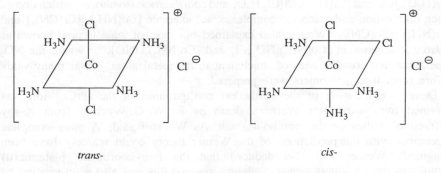

trans-   *cis-*

Figure 12.10 Geometric isomers of $[Co(Cl)_2(NH_3)_4]Cl$

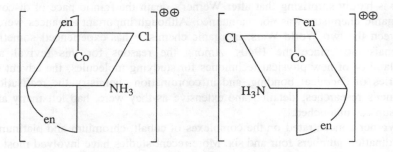

Figure 12.11 Optical isomers of *cis*-$[Co(Cl)(NH_3)(en)_2]^{2+}$

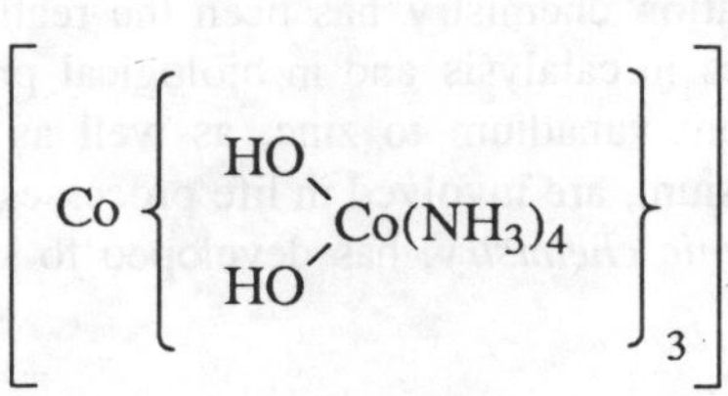

$$\left[Co \left\{ \begin{array}{c} HO \\ HO \end{array} Co(NH_3)_4 \right\}_3 \right] Br_6$$

Figure 12.12 The first carbon–free coordination compound to be resolved into optical isomers by Werner in 1914

Werner's researches provided an overwhelming body of evidence that the hexacoordinate complexes must be octahedral. By counting the number of stereoisomers observed when a central ion is joined to various numbers of monodentate and bidentate ligands in hexacoordinate complexes, Werner demonstrated that the octahedral configuration was the only one consistent with all the observations.

Werner also described and named the various other types of isomerism observed in coordination compounds. These are *ionisation isomerism*, displayed by $[Co(NH_3)_5Cl]SO_4$ and $[Co(NH_3)_5SO_4]Cl$; *hydrate isomerism*, for example $[Cr(H_2O)_6]Cl_3$ and $[Cr(H_2O)_5Cl]Cl_2.H_2O$; and *coordination isomerism*, which occurs when both cation and anion are complexes, for example $[Co(NH_3)_6][Cr(CN)_6]$ and $[Cr(NH_3)_6][Co(CN)_6]$. Werner also explained one case of what is now known as *linkage isomerism* in $[Co(NH_3)_5NO_2]Cl_2$ and $[Co(NH_3)_5ONO]Cl_2$, where the NO_2 ligand has two possible sites of attachment to the metal atom. It was many years before other linkage isomers were prepared.

Direct confirmation of the octahedral configuration of the $[PtCl_6]^{2-}$ ion was obtained two years after Werner's death by R. W. G. Wyckoff from X-ray diffraction studies on the ammonium salt. As Wyckoff said, 'A more complete agreement with the predictions of the Werner theory could scarcely have been imagined'. Werner had also deduced that the four-coordinate platinum(II) complexes have a square planar configuration, and this was also soon verified by the same technique.

It is hardly surprising that after Werner's death the frantic pace of discovery in inorganic chemistry was not maintained. Although important advances were made between the two World Wars, inorganic chemistry has experienced something of a renaissance since the 1940s. Among the reasons for this revival are the availability of new physical techniques for studying molecules, the advent of new theories of chemical bonding, and in coordination chemistry the realisation that Werner's researches, detailed and extensive as they were, had left many areas of the subject untouched.

Werner concentrated on the complexes of cobalt, chromium and platinum, with coordination numbers four and six. More recent studies have involved most metals of the periodic table, and complexes with coordination numbers from two to twelve or even higher have been investigated. An additional spur to the recent rapid development of coordination chemistry has been the realisation of the important role played by complexes in catalysis and in biological processes. All the metals in the periodic table from vanadium to zinc, as well as several others such as molybdenum and magnesium, are involved in life processes. A thriving new branch of the subject, *bioinorganic chemistry*, has developed to study these complexes.

ORGANOMETALLIC CHEMISTRY

The field of organometallic chemistry, concerned with compounds involving metal-carbon bonds, has also expanded enormously since the 1940s. Mention has already been made of Grignard reagents (Chapter 10) and the zinc alkyls discovered by Frankland (Chapter 8). Predating Frankland's work was the discovery in 1827 by William Christopher Zeise (1789–1847) of the salt which was for many years named after him and is now known as potassium trichloro-(ethylene)platinate(II) monohydrate (Figure 12.13). Many other platinum–alkene compounds were prepared, but their structure and bonding were not understood until after the first sandwich compound, ferrocene or bis(cyclopentadienyl)iron(II), was obtained by accident by Peter Pauson and Thomas Kealy in 1951 in the United States (Figure 12.14). The compound had actually been prepared three years earlier by Samuel Miller, John Tebboth and John Tremaine of the British Oxygen Co Ltd, but their paper was not published until 1952. Many other *metallocenes* are now known.

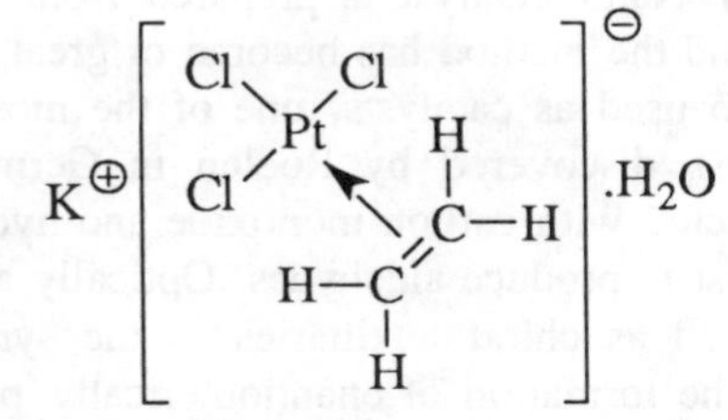

Figure 12.13 Potassium trichloro(ethylene)platinate(II) monohydrate (Zeise's salt)

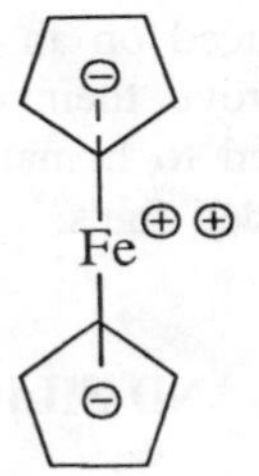

Figure 12.14 Ferrocene

An extremely important class of organometallic compounds was discovered as a result of a chance observation by the industrial chemist Ludwig Mond (1839–1900). Of German extraction, Mond settled in England, and with John Brunner introduced the Solvay process for the manufacture of alkali. Around 1890 he noticed that nickel valves were corroded by hot gases containing carbon monoxide, and with Carl Langer he isolated nickel carbonyl, a colourless volatile liquid, $Ni(CO)_4$. Mond devised a process for purifying nickel based on the volatility of nickel carbonyl, and the Mond Nickel Company, which he founded, built a plant at Clydach in Wales for the production of the metal.

Soon other metal carbonyls were discovered such as $Fe(CO)_5$, $Fe_2(CO)_9$ and $Mn_2(CO)_{10}$. The structure of the carbonyls containing two metal atoms posed a problem until Powell, using X–ray diffraction, showed that in the case of $Fe_2(CO)_9$ the iron atoms are linked by three carbonyl bridges, but in the case of $Mn_2(CO)_{10}$ there is direct metal–metal bonding. It is now possible to synthesise carbonyls and other compounds with many more metal atoms in the molecule or ion, and $[H_2Rh_{13}(CO)_{24}]^{3-}$ was the first species to be prepared in which the central metal atom was surrounded entirely by other metal atoms. Compounds containing several metal atoms joined together are called *metal clusters*, and it is now realised that metal–metal bonding is an important feature of the chemistry of second– and third–row transition elements.

Organometallic and coordination compounds are now finding many uses. In the field of catalysis, one of the most important advances was made in 1953 by Karl Ziegler (1898–1973) when he discovered a method of polymerising ethene at low temperature and pressure. Over the following six years the method was extended to the polymerisation of other alkenes, especially propene, by Guilio Natta (1903–1979). A typical Ziegler–Natta catalyst is prepared from triethylaluminium and titanium tetrachloride, and the method has become of great industrial importance. Metal carbonyls are also used as catalysts, one of the most important examples being in the oxo process discovered by Roelen in Germany in 1938. In this reaction, alkenes are reacted with carbon monoxide and hydrogen in the presence of cobalt carbonyl catalyst to produce aldehydes. Optically active transition–metal complexes are being used as chiral auxiliaries in the synthesis of asymmetric molecules, resulting in the formation of enantiomerically pure products.

Some coordination compounds have found application in the treatment of disease; for example, gold compounds are used in arthritis therapy and platinum compounds are used in the treatment of some cancers. One organometallic compound which has been produced on an enormous scale is tetraethyllead for addition to motor fuels to improve their antiknock properties. An increasing appreciation of the dangers posed to human health by lead compounds is now leading to the phasing out of leaded fuels.

THE HYDRIDES OF BORON AND SILICON

The rapid growth of organic chemistry after 1860 and Mendeleev's periodic classification of 1869 naturally raised the question as to whether there is an extensive chemistry of silicon analogous to that of carbon. A mixture of gaseous silicon hydrides had already been prepared by Wöhler in 1858. He first obtained magnesium silicide, Mg_2Si, by heating the elements in a crucible, and he then treated the product with hydrochloric acid. However, since the silicon hydrides are spontaneously inflammable in air, little progress was made in their investigation until shortly after World War I. At that time Alfred Stock (1876–1946) studied the silicon hydrides using the high–vacuum apparatus he had developed to study air–sensitive compounds. By fractional distillation he separated a series of silicon hydrides analogous to the alkanes in formula, but much more reactive. Attempts to prepare silicon analogues of benzene and the alkenes were unsuccessful.

Before World War I, Stock had used his techniques to investigate the hydrides of boron. Treatment of magnesium boride with hydrochloric acid yielded a mixture which on fractionation gave the compounds B_4H_{10}, B_5H_9, B_6H_{10} and $B_{10}H_{14}$. Stock found that, when B_4H_{10} is heated to 100°C, diborane (B_2H_6) is formed. This molecule has been the subject of intense study, as it is *electron–deficient* in the sense that a conventional structure in terms of electron pair bonds cannot be formulated. Electron diffraction studies have shown the two boron atoms to be linked by two hydrogen bridges. It is believed that each B—H—B bridge bond consists of only two electrons, while the other B—H bonds are of the conventional electron pair type (Figure 12.15). Many other electron–deficient compounds are now known.

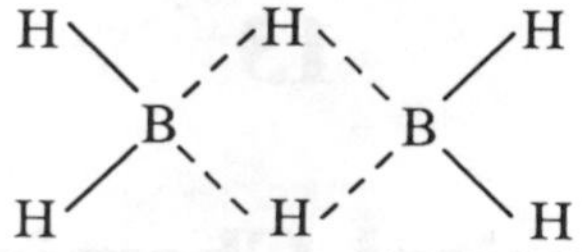

Figure 12.15 Diborane

Stock was also interested in the reform of the nomenclature of inorganic chemistry, and many of his suggestions have been adopted. In the Stock system ferrous and ferric compounds become iron(II) and iron(III) compounds respectively.

RARE–GAS COMPOUNDS

A sensation was created in 1962 when Neil Bartlett (*b*. 1932) of the University of British Columbia announced the preparation of a compound of xenon. Soon after the discovery of the rare gases at the end of the nineteenth century (Chapter 9), many attempts were made to induce them to form compounds. Since all efforts were unsuccessful, it became accepted that these elements were chemically inert. This lack of reactivity was subsequently explained in terms of electronic configuration and became a cornerstone of chemical theory. However, in 1933, Linus Pauling had predicted, largely on the basis of ionic radii, that KrF_6, XeF_6, xenic acid (H_4XeO_6) and salts of xenic acid should exist. Bartlett was investigating the reactions of fluorine with platinum and its compounds in glass and silica apparatus. He accidentally obtained the compound $O_2^+PtF_6^-$ from the reaction of platinum hexafluoride with oxygen adsorbed on the walls of the container. Since the first ionisation energy of molecular oxygen is almost identical to that of xenon, he attempted to react the latter gas with platinum hexafluoride. A yellow solid was produced at room temperature, which was formulated as $XePtF_6$. Since then it has been demonstrated that xenon has a rich chemistry, and some fluorides of krypton are also known. All the compounds predicted by Pauling have been synthesised.

13

Physical Chemistry

We have seen that from the middle of the nineteenth century organic chemistry emerged as a separate discipline in the sense that some chemists devoted all their efforts to this branch of the subject. The date of birth of physical chemistry is often quoted as 1887, for in that year there appeared the first edition of the *Zeitschrift für physikalische Chemie, Stoichiometrie und Verwandtschaftslehre*. The principal editors of the new journal were Friedrich Wilhelm Ostwald (1853–1932) and Jacobus Henricus van't Hoff (1852–1911), and they, along with Svante August Arrhenius (1859–1927), were to champion the new discipline in the face of sometimes scornful attacks from organic chemists. While the appearance of the *Zeitschrift* provides a clear indication that physical chemistry was emerging as a distinct subdivision of the subject, many of the major themes of physical chemistry have their origin before 1887.

THE NATURE OF HEAT

Not surprisingly, people had speculated on the nature of heat since ancient times. Fire was one of the four classical elements and hotness one of the four primary qualities. At the time of the scientific revolution there was a widespread belief that heat was a substance, and some held the view that it was composed of atoms. Francis Bacon, Robert Boyle and Robert Hooke adopted a particulate explanation, and Hooke held the view that a body became hot because of the motion of the particles of which it is composed.

During the eighteenth century Joseph Black, who laid the foundations of calorimetry, preferred an alternative explanation that heat is a fluid. A fluid can be absorbed by bodies or squeezed out of them, and can flow from one place to another, and thus the fluid theory seemed to provide a better explanation of the properties of heat. Black, when he introduced the concept of latent heat, imagined the melting of ice to be a combination between ice and heat, and likewise water combined with more heat when it boiled. Lavoisier embraced the fluid theory and listed heat in his table of the elements, naming it *caloric*. He imagined a gaseous element such as oxygen to be a compound of caloric with what he termed the elementary base of the gas.

The first direct experimental evidence on the nature of heat was provided in 1798 by Benjamin Thompson, Count Rumford (1753–1814). He noticed that a large quantity of heat appeared to be generated when a cannon was being bored. He devised experiments in which a block of iron which was being bored was surrounded by a box full of water. The borer was connected to a lathe, which was turned by two horses. Enough heat was generated to boil the water, and there was no apparent change in the iron itself. Rumford found that the amount of heat generated depended on the work done by the horses, and that there was no limit to the amount of heat which could be produced in this way. Rumford concluded '... it appears to me to be extremely difficult, if not quite impossible, to form any distinct idea of anything being excited or communicated in the manner which Heat was communicated in these experiments unless it be MOTION'.

In the ensuing years, Rumford continued to argue in favour of the dynamical theory of heat. He pointed out that caloric appeared to be weightless, which was hardly consistent with it being a material fluid. However, most of his contemporaries were unconvinced. Dalton depicted the atoms of gases as being surrounded by a sphere of caloric, which was responsible for the repulsion between the particles.

It was not until the 1840s that the caloric theory was finally overthrown. At that time the concept of energy was beginning to emerge. The main contributors were the German Julius Robert von Mayer (1814–1878) and the Englishman James Prescott Joule (1818–1889). Mayer found that there was a numerical equivalence between the mechanical work done and the heat produced, and he subsequently measured a value for the mechanical equivalent of heat. Mayer's work went largely unnoticed at the time. Joule, who had been a pupil of Dalton, studied the heating effect produced by the passage of an electric current, and he then replaced the cells by a dynamo, and reasoned that the source of the heat was the mechanical work expended in turning the dynamo. By experiments such as these he demonstrated the equivalence of heat and other forms of energy. He established the principle of conservation of energy, he firmly advocated the mechanical theory of heat and he made accurate determinations of the mechanical equivalent of heat.

THERMOCHEMISTRY

Black not only introduced the concept of latent heat, but also showed that a quantity of heat could be estimated by measuring the amount of ice it caused to melt. Lavoisier and Laplace employed this principle in their *ice calorimeter* (Figure 13.1). Between 1782 and 1784 Lavoisier and Laplace used this apparatus to measure the quantity of heat evolved in a number of chemical changes and thus laid the foundations of thermochemistry.

Lavoisier and Laplace also conducted experiments from which they concluded that respiration is a kind of combustion. Lavoisier had earlier shown that in respiration oxygen is converted into fixed air (carbon dioxide). He suspected that an animal is able to maintain its body temperature above its surroundings by the release of heat in the lungs during the process, which he believed to be analogous to the evolution of heat occurring in the conversion of oxygen to fixed air when charcoal was burned. To test this hypothesis Lavoisier and Laplace placed a guinea pig in the calorimeter for 10 hours and found that 13 ounces of ice were melted, but the guinea pig's body temperature was almost unchanged. They then measured how much fixed air was produced by the guinea pig in 10 hours. They measured the amount of fixed air produced in the combustion of a weighed quantity of charcoal, and finally they determined the amount of ice melted in the calorimeter when another weighed quantity of charcoal was burned. They calculated that 10½ ounces of ice would be melted when enough charcoal was burned to produce the same quantity of fixed air as the guinea pig produced in 10 hours.

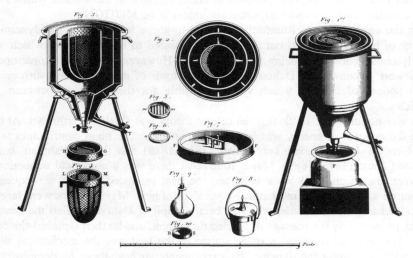

Figure 13.1 The ice calorimeter of Lavoisier and Laplace

In view of the likely errors involved, Lavoisier and Laplace concluded that the quantities of heat released in combustion and respiration were the same. They stated that '... respiration is a kind of combustion, admittedly very slow, but otherwise exactly similar to that of charcoal'. They believed that heat appeared as a result of the liberation of caloric from the oxygen as it combined with carbon from either food or charcoal. Until this time it had been the belief that the purpose of respiration was to cool the blood in the lungs.

Shortly before his death, Lavoisier resumed his experiments on respiration, this time with human subjects (Figure 13.2). He found that the consumption of oxygen increased during digestion, at low temperature and during exercise.

Little further thermochemical work was performed until the 1830s, when Germain Henri Hess (1802–1850) embarked upon an extensive series of experimental investigations. As a result of his work he was able to formulate in 1840 the law named after him, often known as the *law of constant heat summation*. This states that the total heat change in a reaction depends only on the initial and final states of the system, and is independent of the path taken. This law is a consequence of the principle of conservation of energy, but the interconvertibility of energy and its conservation was not properly established at the time. Hess's law continues to find wide application in the estimation of heats of reaction in cases where direct measurement is difficult or impossible.

Figure 13.2 One of Lavoisier's experiments on respiration with a human subject. This sketch is by Madame Lavoisier, who is portrayed taking notes on the experiment

Further thermochemical measurements were made in the 1850s and 1860s by the Danish chemist Julius Thomsen (1826–1909) and the Frenchman Marcelin Berthelot, who had earlier made such important contributions to organic synthesis (Chapter 10). It was Berthelot who introduced the terms *endothermic* and *exothermic* and invented the bomb calorimeter for the accurate determination of heats of combustion. Berthelot suggested that all spontaneous reactions occur with the evolution of heat, and that the reaction which actually occurs in a given situation is the one which is accompanied by the greatest evolution of heat. These conclusions are erroneous, and it was not until the new discipline of thermodynamics was more fully developed that the criteria for spontaneous chemical change were properly understood.

GASES AND THE KINETIC THEORY

The first experiments on the physical properties of gases were performed by Robert Boyle (Chapter 3) when he investigated the relationship between pressure and volume for a fixed quantity of gas. Little significant work was then done until 1802, when Gay–Lussac investigated the expansion of gases as temperature is increased. He measured the volume occupied by a certain mass of gas at the melting point of ice and the boiling point of water. He found that over this temperature range all gases expand by the same fraction of their volume at the melting point of ice. This pointed to a temperature at which all gases would occupy zero volume. This absolute zero was subsequently estimated to be −273°C. At the same time, Dalton carried out similar experiments, but Gay–Lussac's work was more careful and precise.

Gay–Lussac referred in his paper to some work which had been performed by Jacques Alexandre César Charles (1746–1823). Charles had concluded that some gases such as oxygen, nitrogen and hydrogen expand by an equal fraction over the same temperature range, but that water–soluble gases such as carbon dioxide and sulphur dioxide do not. This latter conclusion is incorrect, but nevertheless the law describing the thermal expansion of gases is usually called Charles's law.

In 1848 William Thomson (1824–1907), later Lord Kelvin, proposed that gas thermometers (in which the expansion of gas is measured as the temperature is increased) be used as the ultimate standards against which other thermometers should be calibrated. This proposal was adopted in 1887, and in consequence any modern experiment to 'verify Charles's law' by determining the volume of a certain quantity of gas at different temperatures is bound to reveal a linear relationship, as that is the basis of the calibration of the thermometer. It is interesting to note that Gay–Lussac himself used no thermometer; he simply surrounded his vessels with melting ice and boiling water.

The genesis of Dalton's law of partial pressures can be traced back to his *Meteorological Observations and Essays* (1793). There, he stated that every gas in a mixture acts as an independent entity and there is no chemical combination between them in, for example, the air. Dalton amplified his ideas in papers published in 1801, but experimental evidence was lacking. It was as a result of further speculations about the nature of air that the chemical atomic theory was born (Chapter 6).

In 1829 Thomas Graham (1805–1869) performed experiments on the diffusion of gases. He discovered the law named after him, that the relative rates of diffusion of gases are inversely proportional to the square roots of their densities. In another paper on diffusion published in 1833, he commented that the law 'is not provided for in the corpuscular philosophy of the day'.

The corpuscular philosophy had been applied to gases before, and Newton had shown that it was possible to explain gas pressure and Boyle's law in terms of the repulsions between particles. Newton's picture was essentially a static one, and the first person to attempt to develop a dynamic, or kinetic, theory of gases was the Swiss physician and mathematician Daniel Bernoulli (1700–1782) in 1738. He assumed that gases consist of an immense number of particles of negligible size in rapid motion exerting no forces on each other except when in collision. Pressure is caused by the particles bombarding the walls of the containing vessel, and he showed that Boyle's law could be derived from this model.

The next significant mathematical treatment of the kinetic theory was provided in 1821 by the Englishman John Herapath (1790–1868). Although his treatment contained some errors, it had a stimulating effect on his successors, Joule and Maxwell. In 1848 Joule diverted his attention from his determinations of the mechanical equivalent of heat to calculate the velocity of hydrogen molecules. His value was 6055 feet per second at 32°F (0°C).

The full mathematical development of the kinetic theory was undertaken in 1857 by Rudolph Julius Emmanuel Clausius (1822–1888) in Germany. By this time the concept of energy had been developed, and it was realised that the temperature of a gas was related to the average kinetic energy of the molecules. Clausius was able to derive all the familiar gas laws from the assumptions of the kinetic theory, and in his description of the chaotic motion of gas molecules he introduced the concept of *mean free path*.

Clausius's paper of 1857 also contained work on the specific heats of gases. Clausius realised that, in polyatomic molecules, not all the heat enegy supplied would result in an increase in the velocity of the molecules; some would be absorbed in increasing their internal energy. Clausius showed that, for monatomic gases, the ratio of the specific heats at constant pressure and constant volume should be 1.67, whereas for polyatomic gases the value should be less, as the molecules will rotate more vigorously when heated. The experimentally determined value of the ratio for elementary gases was always found to be less than 1.67, and on this basis Clausius argued that they might consist of diatomic molecules. This piece of physical evidence supporting Avogadro's concept of constituent molecules (Chapter 6) was largely ignored by chemists, and it was not until Cannizzaro argued the case on chemical grounds at Karlsruhe in 1860 that Avogadro's views began to gain acceptance.

Clausius made the simplification of imagining that all the molecules in a sample of gas possessed the same velocity. James Clerk Maxwell (1831–1879) realised that there was a wide distribution of molecular velocities, and in 1859 he applied statistical techniques to the problem. He showed how the various possible velocities are distributed between the molecules and he showed how an average velocity could be calculated. Later, in 1871, Ludwig Eduard Boltzmann (1844–1906) succeeded in generalising Maxwell's distribution law to cases where there are forces (such as gravity) acting on the molecules.

Maxwell derived an equation which linked the viscosity of a gas to the density, the mean free path and the average velocity of the molecules. From measurements of viscosity made with the help of his wife he estimated the mean free path of oxygen molecules at 0°C to be 5.6×10^{-6} cm (modern value 6.6×10^{-6} cm). Maxwell was also able to estimate the mean free path from diffusion experiments, and the two values were in satisfactory agreement.

With the mean free path thus calculated, it was possible to obtain a value for the number of molecules in unit volume, provided the diameter of the molecule was known. A method of estimating molecular diameters was suggested in 1865 by Johann Joseph Loschmidt (1821–1895), and the value he calculated for the number of molecules per cubic centimetre is sometimes called the *Loschmidt number*. It is now known that Loschmidt's value is too small by a factor of 30. This was due to Loschmidt using Mayer's value for the mean free path. Although Mayer's work was more recent than Maxwell's, his value for the mean free path was less accurate.

Although it had been shown that the gas laws could be derived from the postulates of the kinetic theory, it was demonstrated by Henri Victor Regnault (1810–1878) in 1852 and by Émile Hilaire Amagat (1840–1915) around 1869 that all gases deviate to some extent from ideal behaviour, especially at high pressures. This suggested that some of the assumptions of the kinetic theory were invalid, and in 1873 Johannes Diderik van der Waals (1837–1923) devised a gas equation intended to represent the behaviour of real gases more closely than the ideal gas equation

$$PV = RT$$

Van der Waals took into account the attractive forces between molecules and the finite volume they occupy. His equation

$$(P + a/V^2)(V - b) = RT$$

is adhered to reasonably well by most gases.

In 1869 Thomas Andrews (1813–1885) studied the behaviour of gases under pressure, and he developed the concept of *critical temperature*, as the temperature above which a gas cannot be liquefied. Although many gases had been liquefied prior to 1869 (notably by Faraday, Chapter 7), some, such as nitrogen, oxygen and hydrogen, had resisted all attempts and were called the *permanent gases*. As a result of Andrews's work it was realised that their critical temperatures are well below ambient temperature. The critical temperatures of oxygen and nitrogen are −118.8°C and −147.1°C, and when they are cooled by allowing them to expand through a small orifice (the Joule–Thomson effect) it is then possible to liquefy them. This was first achieved on a small scale for air by Raoul Pierre Pictet (1847–1929) and Louis Paul Cailletet (1832–1919), and on a larger scale around 1884 by a number of workers including James Dewar (1842–1923) at the Royal Institution in London. Dewar invented the vacuum flask to insulate the air he had liquefied.

The last 30 years of the nineteenth century was a time in which fierce debates were occurring over the existence or otherwise of atoms. The organic chemists were working away in their laboratories developing new synthetic methods and producing new compounds. Most took the existence of atoms for granted, but atomism remained a useful hypothesis for which there was no direct experimental evidence. Around 1890 Ostwald started to advocate a chemistry based entirely on energetics rather than the kinetic atomic/molecular theory. Ostwald maintained that energy could fill space in an uneven or periodic manner, and that 'matter' is only a complex of energies found together at the same place. Ostwald's thinking was very much in the tradition of Faraday and Boscovic (Chapter 7).

Experimental evidence for the atomic theory was not long in coming. Early in the twentieth century J. J. Thomson identified cathode rays as particles of electricity, Rutherford and Soddy explained radioactivity in atomic terms, and Crookes's spinthariscope enabled individual atomic disintegrations to be detected (Chapter 11). In 1905 Albert Einstein (1879–1955) showed that the Brownian movement of minute suspended particles could be explained in terms of molecular bombardment and he developed a mathematical treatment of the phenomenon. Between 1909 and 1911 Jean Baptiste Perrin (1870–1942) performed experiments on suspended pollen grains in liquids and gases, and was able to estimate a value for the Avogadro number in several ways including measuring the net displacement of a grain in a certain time and the distribution of the grains under gravity. Perrin's value for the Avogadro number agreed well with that obtained by Rutherford and Boltwood in 1911. They counted how many alpha particles a piece of radium emitted per second, and they also measured the volume of helium produced by the radium over a longer period. On the assumption that all the helium atoms had originated from the alpha particles, they calculated the number of helium atoms in the molar volume. In the face of this evidence, opposition to the atomic theory finally collapsed, and even Ostwald announced his conversion to atomism.

REACTION KINETICS AND CHEMICAL EQUILIBRIA

Most of the early studies on reaction kinetics and chemical equilibria had as their aim the quantification of the forces of chemical affinity. Ever since Newton introduced the concept of universal gravitation, those who had studied chemical changes had imagined that attractive forces must exist between the particles of reacting substances.

Probably the first kinetic study was that done by Carl Wenzel (1740–1793) in 1777. He proposed that metals could be arranged in order of affinity for an acid by measuring how long it took for a certain quantity of each metal to dissolve in the acid. He also varied the concentration of the acid reacting with a given metal, and found that when the concentration was halved the piece of metal took twice as long to dissolve.

It had long been realised that many chemical reactions did not go to completion, and it was thought that the measurement of the extent of the reaction might provide a measure of the affinity of the reactants. Many investigations were performed, mostly on heterogeneous systems, and some revealed that the extent of the reaction depended on the amount of one of the reactants present. Thus Heinrich Rose (1795–1864), in a series of investigations in the 1840s and 1850s, studied reactions such as that between solid barium sulphate and a boiling solution of sodium carbonate, resulting in the partial formation of solid barium carbonate and sodium sulphate solution. He found that when equivalent amounts of the two reactants were used, the conversion to products was less than 10 per cent, but if a fifteen–fold excess of sodium carbonate was used, all the barium sulphate appeared to be converted to barium carbonate.

The results of the first thorough kinetic study were published in 1850 by Ludwig Ferdinand Wilhelmy (1812–1864). He investigated the hydrolysis of sucrose in the presence of acid. This reaction could be followed with ease, since the optical rotation of the solution decreased and then changed its sign as the sucrose was hydrolysed (or *inverted*) to glucose and fructose. Wilhelmy showed that the rate of decrease of the sucrose concentration at a particular time was equal to *MZS*, where *M* is a constant, *Z* is the sucrose concentration at that time, and *S* is the acid concentration, which he found remained constant during the reaction.

Wilhelmy's work remained largely unnoticed until Ostwald drew attention to it over 30 years later. In the meantime, three pairs of workers had performed important work on reaction kinetics and chemical equilibria. They were the Frenchmen Berthelot and Léon Péan de Saint–Gilles (1832–1863), the Norwegians Cato Maximilian Guldberg (1836–1902) and Peter Waage (1833–1900), and the Englishmen Augustus Vernon Harcourt (1834–1919) and William Esson (1839–1916).

In 1862, Berthelot and Saint–Gilles studied the homogeneous system of acetic acid and ethanol reacting to produce ethyl acetate and water. They found that the rate of the reaction was proportional to the amounts of reacting substances and inversely proportional to the volume of the system. They found that the same equilibrium position was obtained whether equivalent amounts of acid and alcohol or of ester and water were mixed initially. They also investigated how much acid was esterified by reacting it with different amounts of alcohol. For example, they found that with equivalent amounts of acid and alcohol present initially, 66.5 per cent of the acid was converted at equilibrium, and this rose to 93.2 per cent when the alcohol was present in twelve–fold excess.

The promising researches of Berthelot and Saint–Gilles were terminated in 1863 by the death of the latter at the age of 31. Guldberg and Waage commenced their investigations in 1861, and in a paper published in 1867 they gave their reasons for undertaking the work: '... we thought that it might be possible to find numerical values for the magnitudes of chemical forces. We also thought that we might find for each element and for each chemical compound certain numbers which would express their relative affinities, as atomic weights express their relative weights'.

Guldberg and Waage studied systems which reach a state of dynamic equilibrium, because they could then equate the forces promoting the forward and backward reactions. They studied processes such as the ester formation and hydrolysis investigated by Berthelot and Saint–Gilles, and the heterogeneous equilibria in systems containing sulphates and carbonates of barium and potassium, which had been investigated by Rose.

Guldberg and Waage stated that, if two substances A and B react together to form two new substances A' and B', then the chemical force acting between A and B is measured by *kpq*, where *k* is a constant which they called the *affinity coefficient* and *p* and *q* are the *active masses* (or concentrations) of the reactants. The expression *kpq* is equal to the mass of A and B which is transformed into A' and B' in unit time. At equilibrium, the chemical force of the forward reaction is equal to that of the backward reaction, and thus *kpq* = *k'p'q'*. These conclusions they called the *Law of Mass Action*.

Guldberg and Waage thus arrived at a conclusion similar to the modern equilibrium law. Guldberd and Waage's work did not become as widely known as the authors had hoped. Their original publication in 1864 was in Norwegian and this was followed by a second account in 1867 in French. Over the next few years other workers, including Ostwald and van't Hoff, derived expressions for special cases of the equilibrium law in ignorance of the work of Guldberg and Waage. In

1879 the two Norwegians published another paper in which they restated their earlier conclusions and demonstrated how they had been confirmed by the more recent work of others.

Almost simultaneously with the appearance of Guldberg and Waage's first paper of 1864, Harcourt and Esson commenced their work on reaction rates. Whereas the French and Norwegian workers had been concerned with both equilibria and reaction rates, Harcourt and Esson chose reactions which were effectively irreversible, and they were therefore able to concentrate on chemical kinetics alone. They made no claim to be measuring affinity or chemical force. They initially studied the reaction between potassium permanganate, oxalic acid and sulphuric acid, but this proved to be rather complicated. They therefore switched to the reaction between 'hydric peroxide and hydric iodide' in acidified solution:

$$H_2O_2 + 2HI \rightarrow 2H_2O + I_2$$

They studied this reaction intermittently for about 30 years, performing thousands of careful experiments which rigorously established the dependence of reaction rate on the concentrations of the reactants. In 1895 they described how the rate of reaction varied with temperature. They found that the rate of reaction approximately doubled for every 10°C rise in temperature, and by extrapolation they concluded that at a temperature of −272.6°C no chemical change would take place.

Towards the end of the nineteenth century several other workers were engaged in a study of kinetics and equilibria, most notably van't Hoff in Amsterdam. In 1884 he reviewed much of the work performed to date by himself and others in his book entitled *Études de dynamique chimique*. In this work van't Hoff discussed the relationship between reaction kinetics and reaction mechanism and the dependence of reaction rate on temperature.

Van't Hoff (Figure 13.3) studied at the Polytechnic School of Delft and the University of Leyden. He then worked with Kekulé in Bonn and with Wurtz in Paris, before returning to the Netherlands in 1874. Shortly after his return he published his theory of the tetrahedral carbon atom (Chapter 10). After a spell as lecturer in physics at the State Veterinary School in Utrecht, he transferred in 1877 to the University of Amsterdam, where he eventually became head of the department of chemistry. It was at Amsterdam that van't Hoff performed his important work on kinetics, equilibria and thermodynamics. In 1896 he moved to Berlin, where he was able to devote all his time to research, but his most creative period was at an end. When the first award of the Nobel Prize for Chemistry was made in 1901, van't Hoff was chosen as the recipient.

In 1889 Arrhenius discussed why a small increase in temperature produced such a dramatic increase in reaction rate. He pointed out that the kinetic theory alone is incapable of providing a satisfactory explanation. In gases the predicted increase in the frequency of collisions between molecules when the temperature was raised was far too small to explain the observed increase in reaction rate, and, as Arrhenius pointed out, the situation is probably similar in solutions.

Figure 13.3 Jacobus Henricus van't Hoff (1852–1911)

As early as 1867 Leopold Pfaundler (1839–1920) had suggested that only those molecules possessing more than a critical energy could undergo a chemical change. Maxwell's work on the distribution of molecular velocities had suggested that the proportion of molecules possessing energy considerably in excess of the average at ordinary temperature increased markedly as the temperature was raised. Arrhenius proposed that only highly energetic molecules were *active* and capable of undergoing reaction. Even at high temperatures the number of active molecules is very small compared to the total number of molecules present. Arrhenius gave the relation between the two rate constants k_0 and k_1 at temperatures T_0 and T_1 as

$$k_1 = k_0 \exp[q(T_1 - T_0)/2T_1T_0]$$

This equation is now usually written in the general form

$$k = A \exp[-E/RT]$$

Van't Hoff's *Études* not only discussed kinetics but also considered equilibria as well. He gave the condition for dynamic equilibrium as being when the rates of the forward and backward reactions are equal, and he derived the familiar expression for equilibrium constant. Although he had been anticipated by Guldberg and Waage, van't Hoff gave a much clearer treatment. Using the data of Berthelot and Saint–Gilles for the esterification of acetic acid, van't Hoff showed their results were consistent with the formula

$$[CH_3COOH][C_2H_5OH] = \tfrac{1}{4}[CH_3COOC_2H_5][H_2O]$$

thus giving an equilibrium constant for the reaction of 4.0.

Van't Hoff recognised that change of temperature always affects equilibrium position. In his *principle of mobile equilibrium,* van't Hoff stated that an increase of temperature would favour the endothermic reaction.

At about the same time as van't Hoff's book appeared, the Frenchman Henri Louis Le Chatelier (1850–1936) stated the principle named after him, that when a change is imposed on a system in dynamic equilibrium, the system will respond in such a way as to tend to reduce the imposed change. Van't Hoff's principle of mobile equilibrium was thus shown to be a special case of Le Chatelier's principle.

REACTION MECHANISMS

One of the most important applications of kinetic investigations in chemistry is the elucidation of reaction mechanisms. The pioneer in this field was Arthur Lapworth (1872–1941). In 1903 he studied the addition of hydrogen cyanide to carbonyl compounds (Figure 13.4). By choosing the yellow–coloured substrate camphor quinone (1,2–diketocamphane) he was able to follow the course of the reaction by observing the disappearance of the yellow colour. He found that acids caused the reaction to proceed more slowly whereas bases accelerated the reaction rate. He concluded from this that the real attacking species was the cyanide ion. He suggested that the carbonyl group was polarised, thus facilitating the attack of the cyanide ion. He proposed that the ion thus formed was then rapidly attacked by a hydrogen ion (Figure 13.5).

$$R_2^{R_1}\!\!\diagup C=O \quad + \quad HCN \quad \longrightarrow \quad R_2—\overset{\overset{\displaystyle R_1}{|}}{\underset{\underset{\displaystyle CN}{|}}{C}}—OH$$

Figure 13.4 The addition of hydrogen cyanide to a carbonyl compound

$$R_2^{R_1}\!\!\diagup C=O \quad + \quad HCN \quad \rightleftharpoons \quad R_2—\overset{\overset{\displaystyle R_1}{|}}{\underset{\underset{\displaystyle CN}{|}}{C}}—O^{\ominus} \quad + \quad H^{\oplus}$$

(fast)

$$R_2—\overset{\overset{\displaystyle R_1}{|}}{\underset{\underset{\displaystyle CN}{|}}{C}}—OH$$

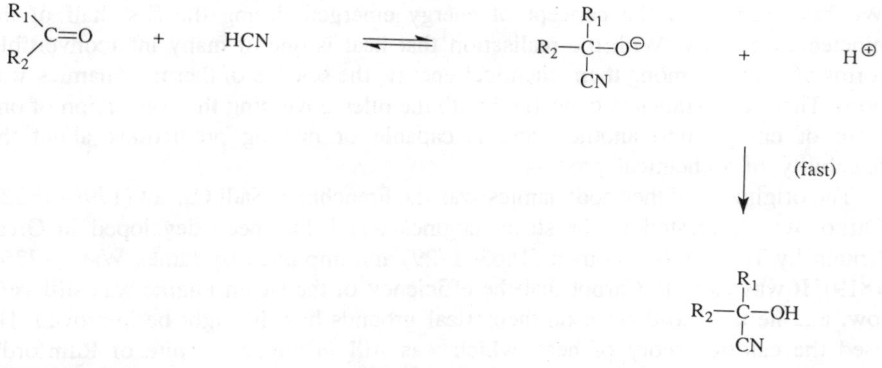

Figure 13.5 Lapworth's proposed mechanism for the addition of hydrogen cyanide to a carbonyl compound

Lapworth was not able to establish the quantitative relationship between reaction rate and cyanide ion concentration, as the rate is very sensitive to low cyanide ion concentrations, which were difficult to determine. However, such quantitative measurements were soon made for other reactions. By determining the order of the reaction with respect to each species it became possible to suggest an equation for the rate–determining step in a reaction sequence.

Lapworth believed that the majority of organic reaction mechanisms involved ions. He proposed that positively and negatively charged reagents be termed *cationoid* and *anionoid* respectively. Lapworth's original work was performed before the electronic theory of valency was introduced by Lewis and Langmuir. A more complete understanding of organic reaction mechanisms became possible when the factors which affect the electronic distribution in organic molecules were understood. The ideas were initially developed by Robert Robinson (1886–1975) and then independently extended and popularised by Christopher Kelk Ingold (1893–1970).

Lewis had discussed the unequal distribution of electrons in a covalent bond. This effect, it was now realised, could be transmitted some way along a chain of atoms, and was termed the *inductive effect*. In 1923 Lowry had proposed that concerted movement of electron pairs in a molecule could occur in the course of a chemical reaction. It was realised in 1926 by both Robinson and Ingold that such displacements make a contribution to the normal state of a molecule, and the phenomenon was named the *mesomeric effect* (Chapter 11). In the same year both Robinson and Ingold started to explain the orientation observed when a monosubstituted benzene was further substituted in terms of the inductive and mesomeric effects. Ingold introduced the terms *electrophilic* and *nucleophilic* for reagents which Lapworth had termed cationoid and anionoid.

CHEMICAL THERMODYNAMICS

We have seen how the concept of energy emerged during the first half of the nineteenth century. With the realisation that heat is one of many interconvertible forms of energy, among them chemical energy, the science of thermodynamics was born. Thermodynamics is concerned with the rules governing the conversion of one form of energy into another, and is capable of making predictions about the feasibility of a chemical process.

The originator of thermodynamics was the Frenchman Sadi Carnot (1796–1832). Carnot was interested in the steam engines which had been developed in Great Britain by Thomas Newcomen (1663–1729) and improved by James Watt (1736–1819). It was clear to Carnot that the efficiency of the steam engine was still very low, and he tried to deduce on theoretical grounds how it might be improved. He used the caloric theory of heat, which was still in vogue in spite of Rumford's work, but this did not affect the validity of his conclusions.

Carnot considered an idealised frictionless steam engine, with heat passing from source to sink through an infinite number of equilibrium states. The pressure–volume changes occurring constitute a *Carnot cycle*. One of Carnot's aims was to find out if steam was the best working substance, and he showed that the efficiency of the engine is not affected by the material used and depends only on the temperature difference between the heat source and sink. Carnot's paper was published in 1824 and aroused little interest at the time. Eventually his conclusions were to be of great importance for both engineers and scientists.

Carnot believed that in an engine heat passes from source to sink unchanged, but Joule, as a result of his studies on different forms of energy, realised that some heat is being converted into mechanical work. Joule was basing his conclusion on the principle of conservation of energy, otherwise known as the first law of thermodynamics. In 1850 Clausius formulated the second law of thermodynamics as 'heat cannot spontaneously pass from a colder to a hotter body'.

In 1865 Clausius introduced the term *entropy* for the heat energy lost or gained by a body divided by its absolute temperature. Another statement of the second law is 'the entropy of an isolated system can only increase or remain constant'. Boltzmann later equated the entropy of a system with its degree of randomness or disorder, and it was Clausius who summarised the first and second laws in the words: 'The energy of the universe is constant; the entropy of the universe tends to a maximum'.

One of those who helped to make Carnot's work more widely known was another Frenchman, Benoit–Pierre–Émile Clapeyron (1799–1864). Clapeyron designed steam locomotives for the railways which were being built in France in the 1830s. In 1834 he published an account of Carnot's work which used differential calculus instead of Carnot's verbal treatment. He derived a relationship between the temperature coefficient of the vapour pressure of a liquid (dP/dT) and the latent heat of vaporisation (L):

$$dP/dT = L/T(V_V - V_L)$$

where ($V_V - V_L$) is the volume change on vaporisation. Clausius recognised that the volume of the liquid (V_L) is small compared to the volume of the vapour (V_V), and by equating the latter to RT/P derived the equation

$$dP/dT = PL/RT^2$$

The first application of these principles to chemical equilibria was achieved in 1869 by August Friedrich Horstmann (1842–1929). He studied the sublimation of ammonium chloride and found that at each temperature a definite vapour pressure is established, and therefore the Clapeyron equation is applicable. It was subsequently discovered that the equation is also applicable to dissociation processes such as that of calcium carbonate (yielding carbon dioxide) and of salt hydrates (yielding water vapour). Replacing L by Q, the energy change for the process, the Clapeyron equation can be written as

$$(1/P)(dP/dT) = Q/RT^2 \quad \text{or} \quad dlnP/dT = Q/RT^2$$

In 1889 van't Hoff extended the equation to chemical reactions in general. Since the dissociation pressure P is a measure of the extent of the reaction, it may be replaced by the equilibrium constant K. Thus

$$\mathrm{d}\ln K/\mathrm{d}T = Q/RT^2$$

This equation, known as the van't Hoff isochore, enables a prediction to be made as to how the equilibrium constant will change when the temperature is altered, and is the basis of van't Hoff's law of mobile equilibrium.

A comprehensive thermodynamic treatment of chemical equilibria had been carried out in the 1870s by Josiah Willard Gibbs (1839–1903). His consideration of heterogeneous equilibria led to the development of the *phase rule*, which enables an evaluation to be made of the number of variables (degrees of freedom) which must be specified in order to define a given heterogeneous system at equilibrium. His work on thermodynamics provided inspiration for later workers, most notably G. N. Lewis at Berkeley, who adapted and applied some of Gibbs's ideas.

Another important concept was developed in 1882 by Hermann von Helmholtz (1821–1894). This was the *free energy* change of a process, which Helmholtz defined as the total external work which can be performed as a result of the energy change accompanying a reaction. On the suggestion of Lewis, the free energy change is now defined as the work available for use, which is the total external work done minus any work done by expansion against atmospheric pressure. This is known as the Gibbs free energy change (or Gibbs function change), and provides a measure of the feasibility of the reaction. Its value can be calculated from the equilibrium constant, and is equal to $-RT\ln K$.

THE CHEMISTRY OF SOLUTIONS

We have seen how Dalton discarded the idea that air is a weak chemical compound of its constituents (Chapter 6). However, a similar view persisted for much of the nineteenth century with regard to solutions of solutes in liquid solvents. It was believed that they were compounds of indefinite composition, with feeble attractive forces between the solvent and solute molecules. It was van't Hoff who developed a useful model of solutions, and he did this as a result of a consideration of the phenomenon of osmosis.

Osmosis, or the net transfer of solvent from a dilute to a more concentrated solution across a suitable membrane, had been known since 1748 when the Abbé Jean Antoine Nollet (1700–1770) noticed that a pig's bladder covering a container of alcoholic solution was ruptured when immersed in water. The phenomenon of osmosis could only be studied properly when sufficiently strong membranes could be produced to withstand the pressure generated. It was Moritz Traube (1826–1894) who discovered in 1867 that a strong membrane could be prepared by precipitating copper ferrocyanide in the walls of a porous pot. He was able to show that for a given solution osmosis occurred until a certain pressure was reached, which is called the *osmotic pressure* of the solution. In 1877 the German botanist

William Pfeffer (1845–1920) experimented on solutions of cane sugar and other substances using Traube's membranes. He found that the osmotic pressure is proportional both to the concentration of solute and to the absolute temperature.

On the basis of these results, van't Hoff in 1883 proposed the theory that osmotic pressure is analogous to gas pressure. Just as the pressure of a gas doubles if its volume is halved (which is equivalent to doubling the concentration of gas molecules), so the osmotic pressure of a solution doubles if the concentration of solute molecules doubles. Applying a thermodynamic treatment to the gaseous model of a solution, van't Hoff derived that the osmotic pressure, like gaseous pressure, should be proportional to absolute temperature, as had been found experimentally by Pfeffer. Van't Hoff also demonstrated that equal molar quantities of different solutes dissolved in the same volume of solvent exert the same osmotic pressure—a law equivalent to Avogadro's law for gases. The analogy with gases was complete when it was shown that dissolved substances exert an osmotic pressure equal to the pressure they would exert if they were gases occupying the same volume. The equation $PV = RT$ could therefore be applied to solutions, where P is the osmotic pressure and R is the universal gas constant.

Osmotic pressure was therefore demonstrated to depend solely on the concentration of solute molecules and to be independent of their nature. Following a proposal of Wilhelm Wundt, Ostwald applied the term *colligative* to such properties. Other properties of solutions which were investigated in the 1880s were the relative lowering of vapour pressure, the elevation of boiling point and the depression of freezing point. These properties were all shown to be colligative in nature by François Marie Raoult (1832–1901), who worked mainly with organic solutes. Raoult demonstrated how the measurement of one of these properties of a solution could be used to determine the molecular weight of the solute.

The van't Hoff theory was found to be inapplicable to solutions of acids, bases and salts, as they yielded osmotic pressures greater than predicted on the basis of their molar concentrations. In these cases the $PV = RT$ equation had to be modified to $PV = iRT$, where i was a factor which varied from one solute to another and increased with dilution. Raoult showed that solutions of these same solutes had abnormally large freezing–point depressions.

In 1884 Arrhenius first proposed a theory which ultimately explained the abnormal behaviour of acids, bases and salts. For his doctorate, Arrhenius studied the conductivity of solutions of these solutes. He found that, when the solutions were diluted, the conductivity decreased, but not by as much as expected from the reduction in solute concentration. Most current explanations of the conductivity of electrolytes were modifications of the theory which von Grotthuss had advanced in 1805 (Chapter 7), namely that the electrolyte particles split up under the influence of the applied field, although Clausius in 1857 had suggested that in solution the molecules of electrolytes are in equilibrium with a small number of charged fragments. Arrhenius proposed that when an electrolyte such as sodium chloride dissolves in water, some of the molecules break up to form free charged particles, and that the percentage of dissociation increases as the solution is diluted. Arrhenius applied Faraday's term *ions* to the charged particles.

Arrhenius (Figure 13.6) grew up in Uppsala in Sweden. He attended the university there and moved to Stockholm for his doctoral studies. His dissertation of 1884 contained the first account of his theory of electrolytic dissociation. Arrhenius's examiners were unimpressed with the theory and he came close to failing his doctorate. However, he sent copies of his thesis to van't Hoff and Ostwald, and they both realised the significance of Arrhenius's ideas. Ostwald offered him a position at his laboratory in Riga, and in consequence the authorities in Uppsala were goaded into appointing him as lecturer in physical chemistry. Between 1886 and 1890 Arrhenius travelled in Europe, visiting the laboratories of Ostwald, Kohlrausch, Boltzmann and van't Hoff. He then returned to Sweden to a lectureship at the Technical High School in Stockholm, where he eventually became rector. Apart from his work on the ionic theory and reaction kinetics, Arrhenius was interested in a wide range of scientific topics. In 1896 he published a paper which for the first time drew attention to the possibility that global warming could result from an increased carbon dioxide concentration in the atmosphere (the greenhouse effect). In 1905 the Swedish Academy of Sciences founded the Nobel Institute for Physical Chemistry, and appointed Arrhenius as its director. He remained in this post until his death.

Figure 13.6 Svante August Arrhenius (1859–1927)

Arrhenius suggested that the abnormally high values of osmotic pressure, freezing–point depression, etc., displayed by these solutions were due to the extra particles they contained on account of the dissociation of the molecules into ions. For a given solution, Arrhenius was able to estimate the degree of dissociation from conductivity measurements and hence calculate the ratio of the actual number of particles present to the expected number if no dissociation occurred. He found remarkably good agreement between this ratio and the ratio of the actual lowering of freezing point to the expected. Some electrolytes, such as sodium chloride, appeared to be highly dissociated in dilute solution, whereas others, such as acetic acid, appeared to be split up to a much smaller extent.

Arrhenius's views were regarded as almost heretical by many of the older chemists of the day. The idea that in solution sodium chloride was largely dissociated into sodium and chlorine was considered absurd. Arrhenius's opponents failed to grasp that the charged ions have very different properties from sodium atoms and chlorine molecules. However, the new theory was championed by Ostwald and van't Hoff, and the first volume of their *Zeitschrift für physikalische Chemie* in 1887 contained an article by Arrhenius in which he gave the first full account of his ideas.

In 1888 Ostwald applied the law of mass action to the equilibrium between an electrolyte and its ions, and derived the *dilution law*

$$K = \alpha^2/(1 - \alpha)V$$

where K is the equilibrium constant, α is the degree of dissociation and V is the dilution. In an extensive investigation of the conductivities of organic acids, Ostwald demonstrated the validity of the dilution law, but it was soon found to be inapplicable to salts and other strong electrolytes.

Ostwald (Figure 13.7) was of German parentage, but was born and brought up in the Latvian capital of Riga. He studied chemistry and physics under German instructors at the University of Dorpat (now Tartu), and he subsequently became professor of chemistry at the Riga Polytechnic Institute. At this time chemistry in Germany itself was almost completely dominated by organic chemistry, and Ostwald subsequently commented that had he been educated in Germany he too would probably have become an organic chemist. As it happened, he developed an interest in trying to measure the forces of chemical affinity. He attempted to correlate the volume changes occurring in acid–base reactions with the affinity of the two reactants, and he studied the kinetics of the acid–catalysed hydrolyses of amides and esters. He employed a range of acids and he correlated the affinity (reactivity) of an acid with its catalytic power. He was therefore in a good position to appreciate Arrhenius's concept of electrolytic dissociation when the latter sent him a copy of his doctoral thesis in 1884. In 1887 Ostwald moved to Leipzig as professor of physical chemistry. For the remainder of his career he championed the ionic theory of Arrhenius against much opposition. He provided additional evidence for the theory, and he developed the theory of acid–base indicators. He resigned from Leipzig in 1905, and in his retirement he worked on the theory of colours, as well as espousing many humanistic, educational and cultural causes.

In the twentieth century, X–ray diffraction experiments have shown that salts exist entirely in the form of ions even in the solid state. The variation in conductivity of solutions of these electrolytes with change in concentration could no longer be explained in terms of increasing degree of dissociation as the dilution was increased. The explanation was provided in 1923 by Peter Debye (1884–1966) and Erich Hückel, who realised that in a moderately concentrated solution of a salt each ion will be surrounded by an atmosphere of oppositely charged ions, which will exert a drag retarding its movement in an electric field. This effect will be more pronounced the more concentrated the solution. Debye and Hückel gave a theoretical treatment which yielded the conclusion that the conductivity of a strong electrolyte should be proportional to the square root of the concentration. Kohlrausch had discovered such a dependence experimentally in 1907. The equation obtained by Debye and Hückel was modified in 1926 by Lars Onsager (1903–1976) to take account of the Brownian motion of the ions.

Figure 13.7 Wilhelm Ostwald (1853–1932)

In 1889 Hermann Walther Nernst (1864–1941) gave a theoretical treatment of the *electrode potential* set up between a metal and a solution of its own ions. He regarded a metal electrode as exerting a *solution pressure*, which is a measure of the tendency of the metal atoms to leave the electrode and enter the solution as ions. At equilibrium, this is balanced by the osmotic pressure of the ions in solution. Nernst derived an equation linking the magnitude of the electrode potential with the ionic concentration and the absolute temperature.

In the same year Nernst also considered the equilibrium between a sparingly soluble salt and its saturated solution. He introduced the concept of *solubility product* and used it to explain precipitation processes.

After Arrhenius had introduced his idea of electrolytic dissociation, it was possible to clarify the nature of acids and alkalis. It had long been known that the essential component of an acid is hydrogen, and Arrhenius proposed that all acids dissociate in water to produce hydrogen ions. Likewise the soluble bases, or alkalis, appeared to be substances that dissociated in water to produce hydroxyl ions. This view of acids and bases received strong support from the observation that the heat energy evolved when an equivalent of a strong acid is neutralised by an equivalent of any strong base is constant. Since strong acids and bases are almost completely dissociated in dilute solution the reaction in all cases is

$$H^+ + OH^- \rightarrow H_2O$$

The fact that pure water has a low electrical conductivity led to the conclusion that it is slightly dissociated into hydrogen and hydroxyl ions. The ionic product for water, $[H^+][OH^-]$, was found to have a value of 1×10^{-14} at 25°C, and the value of the hydrogen ion concentration was found to vary between the limits of approximately 1 M for a very acidic solution and 10^{-14} M for a very alkaline solution. While the value of the hydrogen ion concentration was clearly capable of providing a scale of acidity and alkalinity, the more convenient measure of pH ($= -\log_{10}[H^+]$) was introduced in 1909 by S. P. L. Sørensen (1868–1939).

The definitions of acids and bases were enlarged by proposals put forward independently in 1923 by J. N. Brønsted (1879–1947) and Thomas Martin Lowry (1874–1936). They regarded acids as donors and bases as acceptors of hydrogen ions. An acid can only function as such if a base is present to accept the hydrogen ion from the acid: acid + base = new acid + new base. Each acid therefore gives rise to a new base, called its *conjugate base*, and likewise every base has its conjugate acid. The Brønsted–Lowry definition allowed hydrogen ion transfer processes occurring in non–aqueous media to be classified as acid–base reactions. The new acid–base concept led to the realisation that solvents such as water could act as bases (accepting a hydrogen ion to form H_3O^+) or as acids (donating a hydrogen ion to form OH^-).

The acid–base concept was widened still further by G. N. Lewis with an idea which he first advanced in 1923 and then extended in 1938. Lewis defined an acid as a substance which can accept a pair of electrons to form a covalent bond, while a base is a substance capable of donating a pair of electrons. This definition encompassed all substances previously classed as acids and bases, as well as adding many new ones. For example, the process

$$Na_2O + SO_3 \rightarrow Na_2SO_4$$

would be considered an acid–base reaction under the Lewis definition, as the acid sulphur trioxide accepts an electron pair from the oxide ion to form the sulphate ion.

PHYSICAL PROPERTIES AND MOLECULAR STRUCTURE

From the 1840s many workers were engaged in attempts to correlate various physical properties of compounds with their atomic compositions, and also with their molecular structures as more of these became known. Among the many properties which were investigated were molecular heat, molecular volume, melting and boiling point, and molecular refractivity. The leader in this field of investigation was Hermann Kopp (1817–1892), who also wrote some influential books on the history of chemistry. Useful correlations emerged; for example, properties such as boiling point and molecular refractivity were found to change in a regular manner as a homologous series is ascended. The fact that none of these investigations led to a major technique for the determination of molecular structure should not blind us to the fact that a great deal of valuable and painstaking work was done.

One physical property which did ultimately provide an important method of structure determination was the absorption of infrared radiation by substances. Infrared radiation was discovered by William Herschel in 1800, who found that it had a greater heating effect than light in the visible region of the spectrum (Figure 13.8). Over the next 25 years there were several attempts to repeat Herschel's experiment, some of which were unsuccessful. A systematic investigation by Thomas Seebeck in 1825 revealed that the 'invisible heating rays' were not observed when certain materials were used for the prism, thus indicating that the thermal radiation was being absorbed by the prism itself.

The first systematic attempt to study the absorption of infrared radiation by different compounds was undertaken by W. de W. Abney and E. R. Festing in 1881. Abney developed a photographic method of recording spectra in the near-infrared region, and he and Festing found that compounds could be grouped on the basis of similarity in their absorption spectra and that this grouping 'agrees on the whole with that adopted by chemists'. Abney and Festing declared: 'It seems to us that the spectra leave as definite characters to read as are to be found in hieroglyphics, and we venture to think that we have given a clue to enable them to be deciphered'.

Over the next 30 years much data were collected on the absorption of infrared radiation. It became abundantly clear that absorptions could be correlated with certain bonds or groups of atoms. *Correlation charts* were drawn up which enabled the structural features present in an unknown molecule to be ascertained. Before World War II infrared spectrometers were individually built and therefore the technique was very much a research tool. During the war there was a huge research effort in the United States to find a substitute for natural rubber, as the source of supply in Malaya was cut off by the Japanese. To assist in this search, the first commercial infrared spectrometers appeared, and since that time infrared spectrometry has been widely used in structure elucidation.

With the advent of the quantum theory it became possible to explain infrared spectra in terms of transitions between vibrational and rotational energy levels, and it was realised that it might be possible to estimate parameters such as bond length from a study of infrared spectra. The first attempts at such calculations were performed by Nernst around 1911.

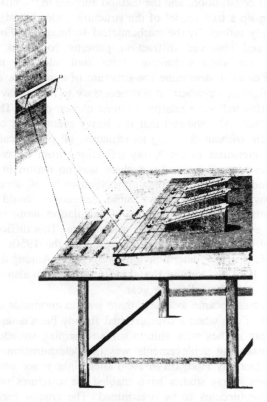

Figure 13.8 Herschel's demonstration of 1800 of the heating effect of infrared radiation. Sunlight passed through a slit in a curtain into a darkened room and through a prism. Thermometers set up on the table beyond the red end of the visible spectrum recorded a temperature rise

Another physical property which has been used in the determination of molecular structure is the diffraction of X–rays by crystals. Reference has been made in Chapter 11 to the experiment of Friedrich and Knipping in 1912 in which they obtained a crude diffraction pattern from a crystal of copper sulphate. The pattern they obtained was too complicated to interpret and they turned to simpler materials. Their experiments were soon repeated by other workers, and within a few months W. L. Bragg had solved the first crystal structures by measuring and comparing the diffraction effects from sodium and potassium chloride.

Over the next twenty years the three–dimensional structures of a number of substances were determined by X–ray diffraction. Among these were diamond and graphite, hexamethylbenzene by Kathleen Lonsdale (1903–1971), and naphthalene and anthracene by J. Monteath Robertson (1900–1989). These results were of enormous theoretical interest, and provided experimental confirmation of the tetrahedral carbon atom and of the planar hexagonal benzene ring.

All the substances investigated in the early days of X–ray diffraction were of known chemical constitution, and the method adopted in the analysis of X–ray data involved setting up a trial model of the structure under investigation. This model was progressively refined by the mathematical technique of *Fourier analysis* until the predicted and observed diffraction patterns coincided. It was, however, impossible to use such techniques with molecules of unknown chemical constitution. In order to determine the structure of a molecule containing 20 or 30 atoms which might be anywhere, it was necessary to find a way of determining the phase of the diffracted wave relative to some chosen origin. This was achieved in 1936 by Robertson, who showed that if a heavy metal atom could be substituted into the molecule without changing its structure, all phases can be determined by comparing the intensities of the X–ray reflections from the two compounds.

As Robertson himself pointed out, there was no reason in principle why any structure, including proteins containing thousands of atoms, could not be determined provided suitable heavy atom derivatives could be prepared. The practical problem was that the amount of calculation work was so large that it would require several human lifetimes to complete. This difficulty was overcome with the advent of electronic digital computers in the 1950s. In the meantime a number of heroic structure determinations were made. Among these were penicillin (Chapter 10) in 1944 by Dorothy Hodgkin (*b*. 1910), who also solved the structure of cholesterol iodide the following year.

When computers became available there was an enormous expansion in X–ray diffraction work. Even when a structure had already been determined by chemical means, the X–ray studies were able to solve remaining stereochemical problems. One of the outstanding achievements was the determination of the structure of vitamin B_{12} by Dorothy Hodgkin in 1956 after eight years' work.

In recent years X–ray studies have enabled the structures of many biologically important macromolecules to be determined. The crucial breakthrough came in 1953 when Max Ferdinand Perutz (*b*. 1914) succeeded in substituting mercury atoms into haemoglobin without changing its structure significantly. The difference in intensities of the reflected X–ray beams from the two types of haemoglobin enabled the phase problem to be solved. The first two protein structures to be determined were myoglobin in 1958 by John Cowdery Kendrew (*b*. 1917), a student of Perutz, and haemoglobin in 1960 by Perutz himself. Their methods have now been applied to several hundred other proteins, including enzymes, antibodies and viral proteins.

Before the work of Perutz and Kendrew had come to fruition, Linus Pauling considered the likely arrangements of peptide chains consistent with hydrogen bond formation and proposed two possible structures. These were the α–helix and the ß–pleated structures. Pauling's suggestions were consistent with the X–ray data then available. It was subsequently found that in many large globular protein molecules the peptide chain consists of some α–helical regions, some ß–structure and some regions where the arrangement is more random.

In 1953 Francis Crick (*b*. 1916) and James D. Watson (*b*. 1928) used an approach similar to Pauling's when they proposed the double helical structure for the DNA molecule (Figure 13.9). On the basis of all the chemical evidence available, they constructed a model and showed that it was consistent with X–ray photographs taken by Rosalind Franklin (1921–1958) and by Maurice Wilkins (*b*. 1916). One particularly important piece of chemical evidence had been obtained by Erwin Chargaff (*b*. 1905). This was that there is an equality between the number of thymine and adenine groups in the molecule, and between the number of cytosine and guanine groups. Watson and Crick proposed the double helical structure in which two coiled repeating sugar–phosphate strands are held together by hydrogen bonding between the complementary pairs of bases adenine–thymine and guanine–cytosine joined to the strands. Crick and Watson commented: 'It has not escaped our notice that the specific pairing we have postulated immediately suggests a possible copying mechanism for the genetic material'.

The understanding of the structures of these macromolecules of living systems has led to the new discipline of *molecular biology*, and the understanding of life processes at the molecular level is one of the most exciting scientific developments of the second half of the twentieth century.

While X–ray diffraction at present provides the ultimate method for the complete determination of the three–dimensional structure of a solid compound, the organic chemist requires a more rapid method for the determination of the molecular constitution of materials, which are often liquids. Infrared spectroscopy still provides a valuable method of qualitative analysis, but more powerful methods of structure determination have been widely employed since the early 1960s. The two most important are *nuclear magnetic resonance* and *mass spectroscopy*.

Nuclear magnetic resonance (NMR) was developed soon after World War II by Felix Bloch (*b*. 1905) and also independently by Edward Mills Purcell (*b*. 1912). The method depends upon the fact that many types of atomic nuclei possess a magnetic moment, which can adopt a number of orientations with respect to a powerful applied magnetic field. Each orientation is a different energy state of the nucleus, and the nucleus can resonate between two states with the absorption of radio waves of the appropriate frequency. The nucleus which has been most widely studied is that of the hydrogen atom, which can adopt just two orientations with respect to the magnetic field. The actual field experienced by the nucleus is affected by the electron density around that particular atom, and therefore the precise resonance condition depends upon the chemical environment of the atom. This effect, known as the *chemical shift*, was first observed by Rex Richards (*b*. 1922) in 1947 when he recorded the existence of three types of hydrogen atom in the NMR spectrum of ethanol. More recently, high–resolution NMR spectrometers have been introduced which enable even more information to be deduced concerning the environment of the resonating atoms in the molecule, and NMR studies on other nuclei, especially ^{13}C, have become common.

Mass spectroscopy employs the method developed by Aston and Dempster (Chapter 11). Before World War II mass spectrographs and mass spectrometers were used almost exclusively to determine the masses and abundances of isotopes. However, it became clear that not only could atoms be ionised in the mass spectrometer, but also that molecules could be made to yield positive molecular ions, and that some of these molecular ions would fragment to yield ions of lower mass. The fragmentation pattern obtained was clearly characteristic of the substance under investigation. This was first utilised in the 1940s, when complex hydrocarbon mixtures were analysed in oil refineries. The mass spectrum of the mixture was compared with the spectra of pure samples of hydrocarbons. With the aid of computers, the composition of a mixture containing 40 components could be determined. Prior to this time the petroleum industry had merely distilled crude oil to produce fractions of different volatility, but when the distilled products were used in chemical processes more careful monitoring of their composition was required.

Figure 13.9 James Watson and Francis Crick in 1953 with their first model of DNA

Mass spectroscopy was soon developed into a powerful method of determining molecular structure. The mass of the molecular ion is now used to give a direct measure of the relative molecular mass of the compound, and the fragmentation pattern enables the structure of the parent molecule to be deduced. Mass spectroscopy is also used in isotopic tracer studies employing non–radioactive isotopes such as ^{15}N and ^{18}O. The technique enables a decision to be made about which compound in a mixture of products contains a heavy isotope.

14

Analytical Chemistry

Analysis of one form or another is an essential and inseparable part of chemistry. The examination of a new material, whether of mineral or organic origin, inevitably involves analytical procedures, as does the investigation of the products of a chemical change. Indeed, analysis of a kind certainly predated both chemistry and alchemy, as the practitioners of the early chemical technologies attempted to identify and quantify the materials with which they were concerned.

In the early history of chemistry, all chemists found themselves performing analyses of one kind or another. In more recent times, analytical procedures have become much more sophisticated, and analytical chemistry has in consequence become a specialised branch of the subject. The skills of the analyst are now utilised not only by other chemists, but also by those carrying out investigations in fields such as forensic science, environmental science and the quality control of all manner of products. Some analytical techniques (for example the quantitative elemental analysis of organic compounds and the investigation of molecular structure by physical methods) have been considered in earlier chapters.

DRY AND WET METHODS OF ANALYSIS

Dry methods of analysis, which generally involve the action of heat on a solid material, were developed to provide methods of assaying metals for purity and ores for their metallic content. Wet methods of analysis, where tests are performed on a solution, had as the principal stimulus to their development the need to analyse mineral waters directly.

One of the earliest dry analytical procedures was to subject gold to the heat of a furnace. Pure gold would remain unaffected, but other metals present would be oxidised or react with the material of the crucible (cupellation). Such analyses could be made quantitative by weighing the gold before and after heating, and balance weights survive which were used in Babylon in 2600 BC.

Mineral ores were usually assayed by procedures which were essentially the same as those used in the extraction of metals but performed on a smaller scale. Agricola described these methods of assay at some length in *De Re Metallica* (1556). The balances used in determining how much metal was obtained from a certain quantity of ore were evidently quite sensitive, and Agricola mentions that the most delicate balances should be housed in a case.

Sometimes a mineral water would be evaporated to dryness and the residue analysed by dry techniques. In a book published in the sixteenth century. it was recommended that the solid should be tasted, the shape of any crystals observed and a portion should be thrown into a fire, when sulphur could be recognised by its characteristic smell and salt by the crackling noise it made. Sometimes the solid was placed on a hot iron in order to observe any changes which occurred on heating.

The use of strong heat to analyse a material was greatly facilitated and extended by the introduction of the blowpipe. This is a narrow tube with which air can be blown into a flame, thereby producing a hotter flame which can be directed at a small area. The blowpipe had been used since antiquity by goldsmiths, and by the seventeenth century it was widely employed in the glass industry. Johann Kunckel (1630–1703) was court alchemist to a German Duke, and in 1679 he published a book on glassmaking in which he suggested that the small blowpipe used by glassworkers would prove a useful piece of apparatus for the chemist. Thereafter it was employed in qualitative analysis, and Bergman and Berzelius both wrote books on the use of the blowpipe.

The usual method of blowpipe analysis was to direct the flame on to a small portion of the material on a charcoal block. The substance under investigation was often mixed with sodium carbonate, borax, or microcosmic salt (sodium ammonium phosphate). When the material was heated alone or with sodium carbonate, it often yielded decomposition products with a characteristic appearance; and borax or microcosmic salt fused to a glass to which the unknown material might impart a characteristic colour. After Wollaston had introduced a method of producing malleable platinum in 1800, a platinum wire was frequently used to support the material in blowpipe analysis, particularly in the production of glassy beads with borax and microcosmic salt.

The second half of the nineteenth century saw blowpipe analysis gradually being replaced by other methods less dependent on the skill of the operator. By this time there had been built up a long tradition of wet methods of analysis, originating in the main from attempts to analyse mineral waters directly. The formation of a black colour when an extract of oak galls was added to a solution containing iron was described in antiquity by Pliny, but the earliest reference to this test in European chemistry is by Paracelsus in 1520. In 1664 Robert Boyle published his *Experimental History of Colours*, which is a landmark in the development of solution analysis. He described how blue plant extracts such as syrup of violets, extract of privet berries, etc., are turned red by acids and green by alkalis. He was the first to notice that all acids and all alkalis have the same effect on a particular vegetable juice. He also noticed that some substances caused no colour change. He classified these as neutral and disposed of a widely held belief that all substances are either acidic or alkaline.

Boyle also discussed precipitation processes. The accepted view was that precipitation occurred when two substances in solution had an aversion for each other, such as when a metal in acidic solution was precipitated by the addition of an alkali. Boyle showed that in some cases a neutral precipitant can be used, such as when sea salt is used to precipitate silver from its solution in aqua fortis (nitric acid). Boyle also washed and dried the precipitate, and showed that it weighed more than the original silver. He concluded from this that a 'coalition' is made between the metal and the precipitant.

From the publication of Boyle's book until the start of the nineteenth century, the number of known precipitating reagents gradually increased. Some of the most important contributions were made by the Swede Torbern Bergman (1735–1784). As a result, wet methods of qualitative analysis assumed an importance equal to and ultimately superior to that of blowpipe analysis. This meant that a mineral had to be got into solution prior to analysis, and this was achieved by grinding it up and treating it with acid or by fusing with alkali prior to extracting with water or acid.

Another consequence of the rise of wet methods was that it opened the way for quantitative analysis by gravimetric means. A precipitate was filtered off, washed, dried and weighed, and from the experimentally determined composition of the precipitate it was therefore possible to work out the percentage of an element (or its oxide) in the mineral. It was the careful application of these gravimetric techniques which resulted in the discovery of several new elements, most notably by Klaproth (Chapter 6).

The increasing range of qualitative and quantitative procedures resulted in 1821 in the publication of the first comprehensive textbook of analytical chemistry by Christian Heinrich Pfaff (1773–1852). Pfaff described each reagent in detail, stating how it should be prepared and giving its analytical applications. Among the newer reagents he described was hydrogen sulphide. Although this had been known for some time, it had only been introduced as a precipitating reagent for metals by Gay–Lussac in 1811.

The next author of note was Heinrich Rose (1795–1864). He published his book in 1829 and it contained for the first time a systematic scheme of qualitative analysis. An improved procedure was published in 1841 in the analytical textbook of Carl Remigius Fresenius (1818–1897). He divided the metals into six analytical groups, and his scheme remains the basis of classical qualitative analysis. Fresenius's book was very successful; by the time of his death it had run to 16 editions and had been translated into many languages. In 1862 Fresenius founded the *Zeitschrift für analytische Chemie*, which was the first chemical journal to specialise in one branch of the subject.

Towards the end of the nineteenth century the range of analytical reagents was extended by synthetic organic substances. The most famous of these is dimethyl-glyoxime, introduced in 1905, which forms a beautiful pink precipitate with nickel ions, and is widely used in both the qualitative and quantitative analysis of that metal.

The quantitative elemental analysis of organic compounds, perfected by Liebig for carbon and hydrogen in 1831 and by Dumas for nitrogen in 1834, has been described in Chapter 8. An alternative method for the determination of nitrogen was devised in 1883 by Johan Kjeldahl (1849–1900) in which the element is converted to ammonia, which is then determined by titration. The standard qualitative method for the detection of the halogens and nitrogen in organic compounds was introduced in 1843 by J. L. Lassaigne, who fused the organic compound with potassium (sodium is generally used today), and then used standard inorganic reactions to detect the presence in the fusion extract of potassium halides and cyanide.

TITRIMETRIC ANALYSIS

A titrimetric method of quantitative analysis is one in which the amount of a reagent required for the completion of a reaction is measured. The earliest such analyses did not involve volume measurements. The first recorded titrimetric procedure was published in 1729 by Claude Joseph Geoffroy (1683–1752), whose elder brother Etienne–François drew up the table of affinities referred to in Chapter 4. Claude Geoffroy devised a procedure for measuring the acid content of vinegars. To a weighed quantity of vinegar, Geoffroy added finely powdered dry potash in small portions until effervescence ceased, and thus the weight of potash used provided a measure of the strength of the vinegar.

It is interesting that Geoffroy's procedure was devised for the quantitative estimation of an article of commerce. In the first 100 years of titrimetric analysis, many methods were developed to enable industrial concerns to assess the quality of the materials they purchased or manufactured. In such circumstances, speed was often more important than accuracy.

A similar method to Geoffroy's was employed in 1756 by Francis Home (1720–1813), who was Professor of Materia Medica at Edinburgh. He helped to establish the chemical industry in Scotland, and he published a book entitled *Experiments on Bleaching*. He described a method for the estimation of potash in which a given weight of the sample was treated with teaspoonfuls of nitric acid until effervescence ceased. The importance of this method is that, although the acid was being measured in terms of teaspoons, it is the first recorded volumetric procedure.

The use of an indicator in an acid–alkali titration is referred to in a monograph published in 1767 by William Lewis entitled *Experiments and Observations on American Potashes with an Easy Method of determining their respective Qualities*. He determined the 'point of saturation' of the alkali by an acid by the 'change of colour produced in certain vegetable juices, or on a paper stained with them'. Lewis advocated that the quantities of both solutions should be determined by weight.

The first volumetric apparatus was designed by François Antoine Henri Descroizilles (1751–1825). He had established a bleaching plant following the discovery by Berthollet of the ability of chlorine to bleach textiles. There was a need for the concentration of the chlorine solutions to be determined, especially as an over–concentrated solution caused the disintegration of the fabric. Descroizilles's procedure was first described by Berthollet in a paper he wrote on bleaching in 1789. The reaction was carried out in a graduated glass tube, which Descroizilles called a *Berthollimeter* (Figure 14.1). A measure of indigo solution was added to the Berthollimeter by means of the small pipette, one pipette measure corresponding to one graduation on the Berthollimeter. The bleaching solution was then run into the Berthollimeter by means of the large pipette until decolorisation occurred. Descroizilles published his own account of his method in 1795, and in 1806 he produced another paper, which was concerned with the estimation of American potash. He used syrup of violets as indicator and added acid to the potash solution using a graduated tube, which he called an *alkalimeter* (Figure 14.2).

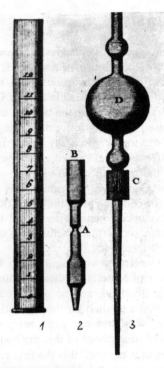

Figure 14.1 Descroizilles's original apparatus for volumetric analysis. The reaction was carried out in the graduated tube on the left, which Descroizilles called a *Berthollimeter*. The other two items are pipettes

The scientist who was responsible for the introduction of titrimetry into mainstream chemistry was Gay–Lussac. He further developed the apparatus used in volumetric analysis and was the first to use the word *burette* (Figure 14.3). In 1832 he introduced precipitation titrations for the estimation of silver. Sodium chloride solution was added to a solution of a silver salt until no further precipitation occurred. This was a time–consuming procedure as near the end-point the precipitate had to be allowed to settle before each new portion of sodium chloride was added.

During the next 30 years many new volumetric methods were devised. One of the most important was the iodometric method introduced by Bunsen in 1853. He determined a wide variety of oxidising substances by causing them to liberate iodine from potassium iodide. He then estimated the iodine produced by running in a solution of sulphurous acid. At about the same time, Karl Leonhard Heinrich Schwarz (1824–1890) recommended sodium thiosulphate for the titration of iodine, and this subsequently became the method of choice. Schwarz published the first book on titrimetric analysis in 1853, and two years later a much more comprehensive work was published by Friedrich Mohr (1806–1879). Mohr collected together many of the earlier methods, and he frequently modified and improved upon them. His book assisted considerably in the popularisation of volumetric analysis.

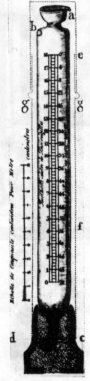

Figure 14.2 Descroizilles's *alkalimeter* of 1806. Acid was poured from this tube into an alkaline solution until neutralisation occurred, as indicated by the colour change of syrup of violets. The flow of acid was controlled by restricting the rate at which air entered through the hole (marked b) by means of a finger. This piece of apparatus was the forerunner of the burette

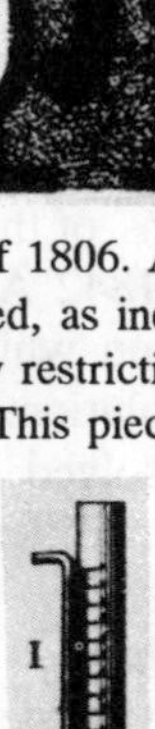

Figure 14.3 Gay–Lussac's burette of 1824. The flow of solution from the jet was controlled by placing the thumb over the larger opening when the burette was tilted

Another important development was the introduction of potassium permanganate as a volumetric reagent by Frédérick Margueritte in 1846. In the same year the French pharmacist Étienne Ossian Henry (1798–1873) devised the first tap burette (Figure 14.4).

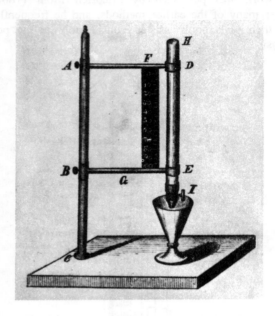

Figure 14.4 The first tap burette devised by Étienne Ossian Henry in 1846. These burettes often leaked through the tap, and Mohr devised a burette in which the tap was replaced by a piece of rubber tubing constricted by a clip. This was very widely used until burette taps became more reliable

In the early days of titrimetric analysis there was a great deal of uncertainty in atomic weights and molecular formulae, and it was not surprising that many chemists expressed the concentrations of their solutions in terms of experimentally determined equivalent weights. In 1842 Andrew Ure (1778–1857) first used the term *normal test liquor* for a solution which contained one equivalent weight in grains dissolved in 1000 grains of solution. Mohr use the term *normal solution* for one equivalent weight in grams dissolved in one litre of solution. Concentrations are still sometimes expressed in normalities, although in recent years the molarity method has become more widely used.

Volumetric apparatus has not changed a great deal since the publication of Mohr's book in 1855. The scope of volumetric analysis has been greatly extended by the use of synthetic indicators, the first of which was phenolphthalein. This was introduced in 1877, and was followed by methyl orange in the following year. This century, many indicators for redox titrations have been developed. Since the 1940s volumetric methods based on complex formation have been introduced, largely as a result of the work of Schwarzenbach. He demonstrated that ethylenediamine–

tetraacetic acid (EDTA) is an excellent reagent for the estimation of many metals, and he also introduced suitable indicators for these titrations.

CHROMATOGRAPHIC METHODS

The first chromatographic separations were achieved in the middle of the nineteenth century, but many years separated the initial observations and the development of a widely used analytical technique. Friedrich Runge (1795–1867) published two books in 1850 and 1855 in which he described how solutions of mixtures of coloured compounds formed concentric rings when spotted on to filter paper. In some of the dyeworks which were established in the second half of the nineteenth century it became the practice to test the quality of a dye mixture by spotting it on to paper or cloth and observing the resulting pattern.

It was the botanist Mikhail Semenovich Tswett (1872–1920) who was responsible for developing adsorption chromatography and demonstrating its application in qualitative analysis. His work stemmed from his observation that certain plant pigments, although readily soluble in petroleum ether, are poorly extracted from the plant by this solvent. His explanation was that the pigments were adsorbed by the plant tissues, and this led him to study the phenomenon of adsorption more closely. In a famous paper of 1906 he described how he 'filtered' a solution of a plant extract through a column packed with powdered calcium carbonate. Coloured bands appeared on the column, which could then be separated by passing down a pure solvent. In Tswett's words: '... the various components of the pigment mixture are separated on the calcium carbonate column, and therefore they can be determined qualitatively and quantitatively'. He demonstrated the power of the method by showing that a leaf extract contained two chlorophylls, four xanthophylls and carotene.

Tswett coined the term *chromatography* (colour writing), but pointed out that the technique could equally well be applied to the separation of colourless substances. Today the term chromatography is applied to any separation technique in which stationary and mobile phases compete for the molecules of the mixture. In spite of Tswett's elegant work it was not until the 1930s that column adsorption chromatography was extensively used by chemists, when it was applied to the separation and purification of a wide variety of natural products.

Partition chromatography originated with the work of Archer John Porter Martin (*b.* 1910) in the 1930s. He devised a technique known as *countercurrent distribution*. The early experiments involved a line of separating funnels containing two immiscible solvents. After shaking the funnels, one of the liquid layers from each funnel was transferred to the next in the line, and the process was repeated. Two solutes travelled at different rates if their partition coefficients for the two phases were different. Subsequently the separating funnels were replaced by special tubes, which meant that the process could be automated.

Partition chromatography was extended by Martin, working in conjunction with R. L. Millington Synge (*b.* 1914). The stationary liquid (water) was held on silica gel, while the mobile phase (chloroform) was allowed to flow over it. The separation of many mixtures was achieved with this system. A very useful form of liquid–liquid partition chromatography was devised by Martin and Synge in 1943 in the form of paper chromatography. Here the stationary phase is water held on the filter paper, and the technique has been widely used for the separation and identification of amino acids.

As well as adsorption and partition, another process which has been employed in chromatographic separations is ion exchange. The ion exchange properties of the zeolite minerals have been known since the 1920s, but it was not until synthetic ion exchangers became available in the 1930s that the technique was widely employed in chemistry. A mixture of ions is added to a column of ion exchange resin. These are then displaced by the ions of the eluting solution, the least strongly held ion appearing first. Ion exchange chromatography was used to separate the fission products (many of them rare earth elements) from the early nuclear reactors in World War II. Ion exchange chromatography is also utilised in amino acid analysers, in which a mixture of amino acids from a protein hydrolysate is separated into its constituents, each of which is estimated quantitatively as it emerges from the column.

In 1953 Martin, in conjunction with A. T. James, developed a partition technique called *gas–liquid chromatography* (GLC). Here the components of a volatile mixture are partitioned between a mobile gas and a stationary liquid, which is contained in a long narrow column surrounded by an oven. Many different detectors have been developed to monitor the gases leaving the column, and the emergence of the components of the mixture is recorded automatically. GLC enables both qualitative and quantitative analyses to be performed on a small quantity of mixture.

Although GLC has proved to be a very successful technique, it is restricted in its application to mixtures of volatile substances. In the 1960s there were attempts to improve column chromatography with a liquid mobile phase so that it could match the performance of GLC for mixtures of involatile compounds in solution. In consequence, apparatus was developed in which the stationary phase is contained in a densely packed column of narrow bore and the liquid mobile phase is pumped through under high pressure. The emerging compounds are usually detected by the absorption of ultraviolet light and recorded automatically. Some of the first instruments to be constructed in this way were designed by C. Horváth in 1966, J. F. K. Huber in 1967 and J. J. Kirkland in 1969. Commercial instruments soon became available, and the technique, known as (high–pressure (or high-performance) liquid chromatography) (HPLC), is now widely employed.

OPTICAL METHODS

It is intuitively obvious that a coloured solute will give a more intensely coloured solution the greater its concentration. The material a chemist wishes to estimate rarely has a pronounced colour, but the practice of forming a coloured product from the substance being measured has a long history. In 1845 Carl Heine (1808–?) published a method to determine the bromide content of mineral waters. A sample was treated with chlorine water, and the free bromine was extracted into a certain volume of ether. The orange colour of the ether solution was compared with a series of standards prepared by treating potassium bromide solutions of known concentration in a similar manner. Other colorimetric processes which were developed around the same time include the estimation of copper by forming the blue tetraamminecopper(II) complex, and the estimation of ammonia by developing an orange–brown colour using Nessler's reagent.

Later in the nineteenth century the colour comparison in such procedures was made more accurate by placing the standard and unknown solutions in special tubes, known as *Nessler tubes*. These were made of glass with a flat base and were housed in a special rack which enabled them to be illuminated from below. The tubes were filled with an equal volume of standard or unknown solution and viewed from above.

As an alternative to comparing the colours of standard and unknown solutions, a number of devices were constructed which enabled the colour of the solution in a tube to be compared with permanently coloured glass standards. These simple instruments were called *Nesslerisers, comparators* or *tintometers* (Figure 14.5).

Figure 14.5 A late nineteenth century tintometer

Early in the twentieth century the colour comparison method was used to measure the hydrogen ion concentration of a solution by comparing the colour developed when an indicator was added both to the solution under test and to a range of buffer solutions. The first such method was devised in 1904 by Hans

Wilhelm Friedenthal (1870–1943), who used fifteen solutions of known hydrogen ion concentration as standards and various indicators, and was able to estimate the hydrogen ion concentration in an unknown solution anywhere in the range of 1 to 10^{-14} molar. A more accurate method, employing a wider range of buffers and indicators, was introduced in 1909 by Sørensen, the inventor of the pH scale. He claimed that the pH value of a solution could be determined to within one–tenth of a pH unit by this method. The colour comparison method of pH determination was simplified by the introduction of indicator mixtures known as *wide–range* or *universal* indicators. These are still commonly used for the approximate determination of pH. In conjunction with the colour comparison method, they showed which buffer solutions and indicators it was appropriate to use. Coloured glass standards were available for pH determinations performed using comparator instruments. Today, the accurate determination of pH is performed using a pH meter.

There were early attempts to develop an improved method of colour comparison, and these depended upon an understanding of the laws governing the absorption of light by a solution. The relationship between the amount of light absorbed and the thickness of the absorbing medium was first enunciated in 1728 by Pierre Bouguer (1698–1758) and restated in 1760 by Johan Heinrich Lambert (1728–1777). They found that successive equal lengths of the medium absorb equal fractions of the light incident upon them. The light intensity therefore decays exponentially as it passes through the medium. Both Bouguer and Lambert passed light through sheets of glass, but in 1852 August Beer (1825–1863) pointed out that the law is also valid for solutions and that the light intensity depends upon solute concentration in the same way as it depends upon path length. Nowadays the name *Beer–Lambert law* is given to the relationship:

$$I = I_0 e^{-\varepsilon c l}$$

where I_0 and I are the light intensities before and after the light has passed through the solution, c is the concentration, l the path length and ε is a constant known as the molar absorption (or extinction) coefficient.

In 1870 the French optician Jules Duboscq (1817–1886) devised a colour comparison instrument (or *colorimeter*). Although not the first of its kind, it was Duboscq's design which was to become standard and is still used occasionally. The two tubes contain the coloured solutions, one of which is the unknown and the other a standard of known concentration. White light is passed through the tubes from below, and, when looking through the eyepiece, half the field is formed by light from one tube and half from the other. The glass plungers in the tubes are moved up or down until the two halves of the field appear to be the same colour. The ratio of the concentrations of the two solutions is then the inverse ratio of the two path lengths (Figure 14.6).

The first instrument which employed a photocell to estimate the light intensity was constructed by Berg in 1911, and from about 1925 photoelectric colorimeters gradually began to replace the Duboscq type. In photoelectric instruments the intensity of the light is measured by a photocell after it has passed through the solution, and this intensity is compared to that obtained when the same beam is passed through a blank. The absorbance (or optical density) of the solution can then be calculated. With the realisation that the Beer–Lambert law is only obeyed when monochromatic radiation is passed through the sample, colorimeters now contain filters, each of which passes a fairly narrow range of wavelengths. The most appropriate filter for the particular analysis is selected beforehand. Since World War II, *spectrophotometers* which operate in both the visible and ultraviolet regions have been developed. By means of prisms or gratings these instruments produce radiation consisting of a much narrower band of wavelengths (typically 2 nm), and utilising the double beam principle they can record the absorption spectrum of a solution over the entire wavelength range of the instrument.

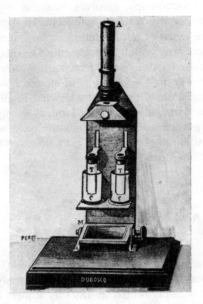

Figure 14.6 Duboscq's colorimeter of 1870

The development of atomic emission spectroscopy by Bunsen and Kirchhoff, and its application as a very sensitive method of qualitative analysis, has been described in Chapter 9. The possibilities of using the technique quantitatively were first discussed by Lockyer in 1874. He realised that, while qualitative analysis rested on the positions of the observed lines, their intensities might form the basis of quantitative analysis. Although attempts were made to measure line intensities by recording them photographically and estimating the blackening of the photographic film, it was not until photoelectric methods of measuring light intensities were introduced by H. G. Lundegardh in 1929 that emission spectro-scopy was widely applied in quantitative analysis.

The principal emission spectroscopic techniques in use today are *flame photometry, inductively coupled plasma spectroscopy* and *spark emission spectroscopy*. The first commercial flame photometers appeared in 1948. The

solution is sprayed into a flame and the emitted light passes through a mono-
chromator or filter specific for the metal under test and its intensity is measured.
The technique is only sufficiently sensitive for the estimation of alkali metals, as
the percentage of atoms of other elements excited in the flame is too low. The
much higher temperature achieved in inductively coupled plasma spectroscopy
(first reported in 1964) gets round this problem, and many more elements can be
analysed by this technique. Spark emission spectroscopy finds application in the
analysis of alloys, especially steels. A spark is struck between poles made of the
steel. The light impinges on a grating and slits are arranged to allow characteristic
lines of the dispersed light to fall on to photomultipliers. The elemental
composition of the sample can then be computed very rapidly.

Atomic absorption spectroscopy involves beaming light of an appropriate
wavelength into a flame into which the sample has been sprayed, and thus contains
an atomic vapour of the metal. The diminution in the intensity of the radiation is
correlated with the concentration of the element. The majority of atoms in the
flame remain in the ground state, so the technique is potentially more sensitive
than flame photometry. The first application of atomic absorption in quantitative
analysis was the determination of mercury by Hewlett in 1930, but it was not until
after A. Walsh introduced the hollow cathode lamp as the light source in 1955 that
the method came into general use. Today, it is the most widely used method of
estimating metals in solution, but is likely to be overtaken in the future by
inductively coupled plasma spectroscopy.

ELECTROMETRIC ANALYSIS

After the work of Nernst on electrode potentials in 1889 (Chapter 13), it became
apparent that EMF measurements could be used in analytical work. The first
potentiometric titration was performed in 1893 by Robert Behrend (1856–1926).
He titrated a solution of mercury(I) nitrate with potassium chloride, thus precipitating mercury(I) chloride. The titration vessel contained a mercury electrode, and was
separated by a porous membrane from another container of mercury(I) nitrate with
a mercury electrode. The EMF of the resulting cell depended on the concentration
of the mercury ions in the titration vessel, and showed a pronounced change in the
region of the end–point.

In 1893, Max Le Blanc (1865–1943) found that a stream of hydrogen gas
flowing over a platinum plate coated with platinum black formed an electrode
which was reversible with respect to hydrogen ions. Using this electrode, Wilhelm
Böttger (1871–1949) was able, in 1897, to monitor an acid–base titration. Böttger
placed one hydrogen electrode in the solution being titrated and another in a
solution of acid or base in contact with it. The end–point was clearly indicated by
a change in the EMF of the cell. In 1895, Olin Tower (1872–?) showed that
electrical contact between two solutions was best achieved by a bridge of
potassium chloride solution.

Potentiometric titration remained something of a curiosity until the 1920s, since when it has been much more widely used. The principle has remained the same as in the early experiments, namely the potential difference is measured between an indicator and a reference electrode.

There were early attempts to determine the pH of a solution by an electrometric method. Sørensen in 1909 used a hydrogen electrode in conjunction with a calomel reference electrode. The hydrogen electrode is inconvenient for routine measurements, and indicator methods were more commonly used for pH determination until the glass electrode was introduced by Duncan Arthur McInnes and Malcolm Dole in 1930. The glass electrode consists of an internal electrode dipping into an electrolyte separated from the solution under test by means of a thin membrane of conducting glass. The possibility of using such an electrode had been suggested by Fritz Haber (1868–1934) and Zygmunt Klemensiewicz (1886–1963) in 1909, but at that time it was not possible to find a glass which was responsive only to hydrogen ions. Dole and McInnes discovered a glass superior to those which had been tried previously, although their electrode was still subject to errors in very alkaline solutions. The modern *pH meter* measures the potential difference between an improved glass electrode and a reference electrode, both of which are usually housed in the same probe.

In addition to the glass electrode responsive to hydrogen ions, other *ion selective electrodes* are now available. These electrodes were first developed around 1964, and E. Pungor was associated with some of the early developments. A great many ions can now be estimated in this way, and among the more widely used electrodes are those responsive to potassium, calcium, fluoride and nitrate ions.

An electrometric technique that assumed considerable importance in the middle decades of the twentieth century was polarography, which was first described in 1922 by Jaroslav Heyrovsky (1890–1967). This technique allows both the identification and estimation of metal ions in low concentration. The solution under investigation is electrolysed in a cell with a cathode which consists of a droplet of mercury at the end of a capillary tube and which increases in size until it is released. A fresh drop then starts to form, and in this way a clean electrode surface is maintained. The anode is a pool of mercury at the bottom of the cell. The voltage is increased steadily and a current–voltage curve is plotted. When the potential difference is large enough to discharge the ion, the current increases sharply, generating a polarographic *wave* on the graph. The *half–wave potential* is characteristic of the ion being discharged, and the height of the wave (the *diffusion current*) is a measure of the concentration of the ion. Polarographic analysis is now being superseded by techniques such as atomic absorption and inductively coupled plasma spectroscopy.

There are many sophisticated modern analytical techniques which cannot be mentioned in this brief survey. Our analytical methods will continue to be extended and improved as increasingly complex substances and mixtures need to be identified, and quantitative estimations of ever smaller concentrations are undertaken. We will never be completely satisfied with our analytical methods, but the range of problems which can be tackled will be continuously widened.

THE SOCIAL HISTORY OF ANALYTICAL CHEMISTRY

In the middle of the nineteenth century the general population had little appreciation of the discoveries that were being made in chemistry or the contribution that chemistry had made to the Industrial Revolution. However, popular attention in Britain was captured by disagreements between chemists called as witnesses in legal cases. Murder trials were of particular interest, especially when the experts were unable to agree if poison had been administered. In reality, analytical chemists were learning by painful experience of the necessity of testing the purity of all reagents before commencing the analysis, and they were also discovering that other species present in the sample could interfere with their tests.

There was clearly a need for analytical methods to be standardised, and this need was given added impetus from various industrial and commercial quarters. The analysis of 'chemical manures' provides an example. Chemical manures were superphosphate fertilisers made by treating ground bones or mineral phosphates with sulphuric acid, but they also incorporated slaughterhouse refuse, blood, scrap leather and various other ingredients. The quality of a manure was generally judged by its potassium and phosphate content, but analysis for these components by gravimetric means was difficult, and different analysts used widely different variations of the same basic method. Depending on the figures which they tended to give, analysts were known as 'high' or 'low'. Standardisation of methods and practice was assisted in Britain by the foundation of the Society of Public Analysts in 1874, which began to publish its own accounts of analytical procedures.

Another area in which analytical chemistry was to come to the notice of the general public in the nineteenth century was the detection of poisonous pigments in confectionery. Among the many substances used were Scheele's green (copper arsenite), chrome yellow (lead chromate) and vermilion (mercury(II) sulphide). Not surprisingly, there were instances of tragedies caused by the use of these compounds. In England in 1848 many people were taken ill and one person died after a public banquet in which the blancmange had been coloured with Scheele's green.

Attention was drawn to the use of these pigments by a number of reports, and in 1860 the first Food and Drugs Act was passed. Further legislation a few years later permitted local authorities to employ public analysts. When investigating suspect confectionery, the analyst would wash the specimen with water to remove the sugar and then subject the residue to simple dry or wet tests. Legislation, the employment of public analysts and the introduction of the new coal–tar dyes resulted in the disappearance of poisonous inorganic chemicals from confectionery. The synthetic organic dyes were initially thought to be perfectly safe food colorants; in our own day, doubts have been raised about some of these substances also. Today, a very large number of analytical chemists are employed in laboratories concerned with public health, forensic investigations and quality control.

It was through the activities of analytical chemists that public awareness of chemistry was increased in the nineteenth century. Towards the end of the century the products of the organic chemical industry began to make a significant impact on peoples' lives. These new substances included not only the synthetic dyes, but also drugs, anaesthetics and antiseptics. More people were educated and employed as chemists, and chemistry became a recognised profession. Some aspects of this wider interaction between chemistry and society will be considered in the next chapter.

15

Chemistry and Society

Previous chapters have described the development of chemistry and the lives of some of those who have shaped the subject. In this concluding chapter we shall consider briefly some aspects of the interaction between chemistry and society.

THE EMERGENCE OF CHEMISTRY AS A PROFESSION

Chemists, like any other group of people, operate within a social context, and as chemistry has developed it has acquired its own organisation within society. There are now a large number of practitioners of chemistry, and chemistry is a recognised profession.

In the days of alchemy, and for many years afterwards, the majority of serious practitioners of chemistry were either medically qualified or were gifted (and usually wealthy) amateurs such as Robert Boyle and Henry Cavendish. Robert Hooke in the seventeenth century was one of the first people to make a living from scientific employment, but the possibility of an individual making a career in any branch of science was rare until the nineteenth century. Chemistry did not become established as a recognised profession until it offered employment to significant numbers, and the opportunities of acquiring a chemical education had been increased.

In the second half of the eighteenth century, chemistry began to be taught as a subject in its own right, and not merely as an adjunct to medicine. One university where this happened was Glasgow, where the first lecturer in chemistry, William Cullen (1710–1790), was appointed in 1747. Cullen and his successors, who included Joseph Black and Thomas Thomson, liberated chemistry from the confines of the medical curriculum. Cullen presented chemistry as a science 'fit for the study of gentlemen', and he stressed the practical utility of the subject, entitling one course 'Chemical lectures and experiments directed chiefly to the arts and manufactures'. Many of the Glasgow lecturers subsequently migrated to Edinburgh, where they followed similar policies. There were similar developments in other European countries. The number of university posts in chemistry in Germany increased considerably in the second half of the eighteenth century, and in eighteenth century France chemistry was taught in its own right at public lectures and at military schools and academies.

In England there was little broadening of the university curriculum, and increased opportunities for studying chemistry arose for another reason. Those who belonged to Nonconformist Christian sects were excluded by law from the universities and the traditional professions, so the Nonconformists established a number of educational institutions of their own known as *dissenting academies*. Joseph Priestley and John Dalton, themselves both Nonconformists, taught at dissenting academies. While one of the aims of these was to train ministers for the various Nonconformist sects, they were free to teach whatever subjects they chose. Thus, while the traditional schools and universities offered a curriculum which consisted largely of classics, some of the dissenting academies introduced the study of science, and it was Priestley who was instrumental in introducing the study of chemistry at Warrington Academy (Chapter 4).

In the early decades of the nineteenth century, a chemical industry was being established to satisfy the needs generated by the Industrial Revolution. The early industrialists were the first significant body of people to make their living out of chemistry. With the steady increase in the number of people being educated in chemistry or making their living from the chemical industry, it was not surprising that those interested in chemistry should form themselves into groups. During the period 1780–1840 several short–lived chemical societies were formed in Britain, and similar developments occurred in mainland Europe and in the United States. Some of the British societies met in coffee houses, while others were more formally constituted, such as the one set up in Edinburgh around 1785 by the students of Joseph Black, and the London Chemical Society founded in 1824 by George Birkbeck (1776–1841).

These societies all faded away in time, but 1841 saw the foundation in London of the precursor of the Royal Society of Chemistry, known originally as the Chemical Society. Today, it is the world's oldest surviving chemical society. The new society brought together academic, manufacturing and consulting chemists from all parts of Britain. It served the academic community by providing a forum at which papers could be presented, and issued a journal in which they could be published. Manufacturing and consulting chemists could discuss the relative merits of various products and processes at meetings of the society, rather than publishing individual pamphlets to expound their views.

The Society provided a means for the professional self–advancement of chemists, particularly the academic chemists, who quickly acquired the dominant position in the running of the Society's affairs. Its first President was Thomas Graham (1805–1869), who was a former student of Thomas Thomson at Glasgow and who had been appointed professor of chemistry at University College in London in 1837.

Chemical societies similar to the one founded in Britain in 1841 were started in France (1855), Germany (1866), Russia (1869) and the United States (1876). The American Chemical Society came into being as a result of a proposal made in 1874 at the centenary celebrations of Priestley's discovery of oxygen. These were held in the house Priestley had built at Northumberland, Pennsylvania.

The Chemical Society founded in Britain continued to prosper, but there was growing dissatisfaction over the lack of recognition of chemists in industry and public service. There was therefore agitation for the formation of a new body to promote the professional interests of chemists in public and industrial employment. This led to the foundation of the Institute of Chemistry in 1877. To qualify for membership, an applicant had to have a university degree or pass the Institute's examination after three years of study at a recognised institution, and, in addition, to achieve full membership, the candidate had to have practised as a chemist for three years.

In 1980, the Chemical Society, the Institute of Chemistry and two other societies which had been formed in the meantime (the Faraday Society (for physical chemistry) and the Society for Analytical Chemistry (formerly the Society of Public Analysts)), merged to form the Royal Society of Chemistry. Early in the twentieth century the American Chemical Society organised itself into divisions representing various specialisms and interests, and was thereby more successful in preventing the proliferation of societies which occurred in Britain.

Chemistry, along with the other sciences, now offers employment to huge numbers of people worldwide. The largest employer of those with chemistry degrees or other professional qualifications is the chemical industry, and the products of this vast industry provide the clearest contemporary example of the interaction between chemistry and society.

THE ORIGINS OF THE CHEMICAL INDUSTRY

So dependent is modern society upon manufactured chemicals that the present period has been called the *chemical age*, although terms such as the electronic age and the nuclear age might be equally appropriate. The huge range of chemicals manufactured today is produced by an industry which is one of the largest in the developed world. Yet some of the substances made in vast quantity are totally unknown to the general public. An example is sulphuric acid, the consumption of which per head of population has often been said to provide a good indication of the industrial health of a nation. Sulphuric acid itself is rarely encountered in everyday life, but it is employed in fertiliser manufacture, the processing of metals, the preparation of dyes and drugs, detergent manufacture and in many other industries.

The chemical industry had its origins in prehistoric times. The extraction of copper, silver, iron, tin and lead from their ores were all performed in ancient times, and the ancients also used salts such as natron (sodium carbonate) for making glass and alum for tanning leather and for mordanting dyes to cloth. The Greeks and the Romans were for the most part content to exploit more fully the processes and materials already known. It was not until the Renaissance that existing chemical processes started to be scaled up and new ones introduced. The Renaissance was marked by a great increase in the demand for metals for both decorative and structural purposes, and there was a consequent increase in the demand for chemicals used in metallurgical processes. Books such as Agricola's

De Re Metallica not only disseminated knowledge about the extraction of metals, but also gave recipes for the production of a number of chemicals.

However, it was the Industrial Revolution, commencing about 1760, which resulted in the establishment of a fully fledged chemical industry. The establishment of a large number of textile mills created a demand for a range of chemicals used in the processing of fibre into finished cloth. These included alkalis, acids, bleaching agents, mordants and dyes.

The traditional source of alkali had been wood ashes, which were extracted with water to produce potash (potassium carbonate). By the Industrial Revolution the forests of Europe had been seriously depleted, and although potash was imported from the New World, demand still exceeded supply. Another source of supply for alkali was the ashes obtained by burning a plant (*Salsola soda*) growing in Spain. The ashes, known as *barilla*, furnished soda ash (sodium carbonate). There was clear need for a method of synthetic alkali production, and, as the Industrial Revolution began to gather momentum, there were a number of attempts to devise a commercially viable method of making alkali from common salt. In 1775 the French Government encouraged these attempts by offering a prize of 100 000 francs to the person who proposed the first satisfactory process. The competition was won in 1790 by Nicholas Leblanc (1742–1806). His process involved a first stage in which salt was treated with sulphuric acid to make sodium sulphate and hydrogen chloride. The sodium sulphate was then roasted with limestone and coal to produce *black ash*, which consisted principally of sodium carbonate and calcium sulphide. The sodium carbonate was extracted from the black ash with water and crystallised.

Leblanc's life ended in tragedy. He was never paid the prize money for his invention, and in 1791 the French revolutionaries confiscated both his plant and his patent rights. Although his property was returned to him by Napoleon, he had no capital to operate his process. By 1806 he was completely destitute, and he committed suicide.

Alkali from the Leblanc process gradually displaced that from vegetable sources, although some consumers were reluctant to switch to the new product. By the middle of the nineteenth century, the Leblanc process reigned supreme. A serious problem associated with the process was the pollution caused by the hydrogen chloride released during the first stage of the process. This was largely alleviated by the invention of William Gossage (1799–1877). Gossage allowed the waste gases to ascend a tower packed with coke down which water was trickling.

In spite of the success of the Leblanc process, it was evident from the outset that its operation involved serious drawbacks. Although the hydrogen chloride pollution problem was largely solved by the Gossage towers, there remained the problem of the residue left behind after the black ash had been extracted with water. The principal constituent of this *alkali waste* was calcium sulphide, but it also contained calcium hydroxide and unchanged coal. There was no alternative but to dump it near the alkali works, where it released hydrogen sulphide when acted upon by acidic rain water. Not only did this cause a serious pollution problem, but also all the sulphur of the original sulphuric acid was wasted.

Although methods were eventually devised to recover the sulphur from the alkali waste, it had been realised from 1811 that there might be a much more elegant alternative process of alkali manufacture. In that year the Frenchman Augustin Jean Fresnel (1788–1827), who is now remembered chiefly for his researches on optics, discovered that when a concentrated solution of salt is saturated with ammonia and then treated with carbon dioxide, sodium bicarbonate is precipitated. This can be filtered off and decomposed by heat to yield sodium carbonate. Ammonium chloride is left behind in solution, and as ammonia was at the time a very expensive material, a way had to be found for recycling it. This was later achieved by treating the ammonium chloride with lime, thus generating ammonia and leaving the harmless waste product calcium chloride. Several manufacturers struggled to introduce the *ammonia–soda process*; the difficulty lay in devising a plant which made the process industrially practicable and involved little loss of ammonia. Real success was first achieved by the Solvay brothers in Belgium in the 1860s, and their process was finally perfected in the 1870s. By 1902 over 90 per cent of the world's soda was being made by the ammonia–soda process, and the method is still in operation today.

Sulphuric acid had been manufactured since the sixteenth century by the distillation of green vitriol, and in the seventeenth century Glauber introduced the method of burning a mixture of sulphur and nitre under a bell jar. The sulphur trioxide formed was absorbed in water to give dilute sulphuric acid. This method of manufacture was much improved by John Roebuck (1718–1794), who at Birmingham in 1746 started to perform the reaction in lead chambers rather than in the fragile glass vessels used previously. The *lead chamber* process was introduced into France in 1766 and into the United States in 1793. The size of the chambers was progressively increased, and by the middle of the nineteenth century they were as large as 56 000 cubic feet. An important improvement in the method was introduced in 1793 by the Frenchmen N. Clement (1779–1844) and C. B. Desormes (1777–1857), who found that the quantity of nitre required could be greatly reduced by admitting air to the chamber. In the nineteenth century the practice of burning the sulphur outside the chamber was introduced, and when the price of sulphur rose the sulphur dioxide was produced by the roasting of pyrites.

The lead chamber process was eventually superseded by the *contact process*, in which sulphur dioxide and oxygen are made to react directly by passage over a catalyst. This method was devised in 1831 by a vinegar manufacturer from Bristol named Peregrine Phillips, but was not exploited commercially on a significant scale until the second half of the nineteenth century, when several factories were founded in Germany. It was at this time that the synthetic dyestuffs industry was growing rapidly in Germany, and it created a demand for fuming sulphuric acid, or *oleum*, which the contact process was capable of producing directly. The lead chamber process produced a more dilute acid suitable for the Leblanc process and for making fertilisers. Phillips used a catalyst of finely divided platinum, but it was later shown that various metal oxides could be used instead. An initial problem with the contact process was that the catalyst ceased to be effective after a while. In 1870 Rudolph Messel (1847–1920), a German chemist working in London, showed that the catalyst was being poisoned by impurities in the reactants, and that the problem was avoided if these were carefully purified.

The expansion of the textile industry also meant that there was much more cloth which needed to be bleached. Traditionally, cloth was bleached by exposure to the sun in bleaching fields; this was a very time–consuming process, especially in northern latitudes. In 1785 Berthollet recommended the use of chlorine in the bleaching of textiles, and chemical bleaching soon displaced the traditional technique. Berthollet dissolved the chlorine in caustic potash solution, but an important improvement was made in 1799 by Charles Tennant (1768–1838) in Glasgow. He absorbed chlorine in lime, thereby making the solid product known as bleaching powder. This became another important product of the chemical industry.

The caustic alkalis (sodium and potassium hydroxides) had been traditionally made by treating the mild alkalis (sodium and potassium carbonates) with lime water. The caustic soda produced by this method was not very pure, and this caused problems in the early 1890s for Hamilton Young Castner (1858–1899) when he tried to develop a process for the manufacture of sodium based on Davy's method of the electrolysis of fused caustic soda. Castner therefore set about devising a method of manufacturing pure caustic soda. It was already known that the electrolysis of brine yielded chlorine at the anode, and hydrogen and caustic soda at the cathode, but there was no known method of separating the caustic soda from the solution in the cell and preventing it reacting with the chlorine. Castner devised a cell with a mercury cathode at which sodium was discharged to form an amalgam with the mercury. The sodium–mercury amalgam reacted with water in a separate compartment of the cell to form caustic soda. The circulation of the mercury was achieved by rocking the cell. Castner found that a somewhat similar process had been devised at about the same time by the Austrian chemist Carl Kellner (1850–1905), and rather than embark on litigation to establish priority the two men reached an agreement. The method was known as the Castner–Kellner process, and it became the standard method of manufacture of both sodium hydroxide and chlorine.

By the end of the nineteenth century it had become clear that a crisis was looming in agriculture. The warning was most forcefully made in 1898 by William Crookes in a speech to the British Association when he pointed out that the vast nitrate deposits in Chile were being exploited at such a rate that they would be exhausted within a generation. Without the application of nitrates to the land, crop yields would fall dramatically and food shortages would be inevitable. There was clearly a need to develop a process to fix atmospheric nitrogen to enable nitrate fertilisers to be made artificially.

It was Fritz Haber (1868–1934) who devised the nitrogen fixation process which was to come into general use. He started work on the reaction between nitrogen and hydrogen in 1904, initially determining the very low equilibrium concentrations of ammonia obtained at atmospheric pressure. With Le Rossignol he then constructed an apparatus that would operate at 150–200 atmospheres pressure, but there was still no catalyst available which would promote rapid reaction at 550°C. Haber eventually used uranium, and after the feasibility of the method had been demonstrated in the laboratory, Carl Bosch achieved the difficult task of developing the process on an industrial scale. The first ammonia plant came into operation in Germany in 1913. The ammonia was then converted into nitric acid, which was used to make fertilisers and explosives. Another method of nitrogen fixation was devised in Norway, in which air was swept through an electric arc. This caused the combination of nitrogen and oxygen, and thus provided an alternative route to nitric acid. However, in spite of Norway's abundant supply of cheap electricity, the process could not compete with that devised by Haber.

Haber (Figure 15.1) was born in Breslau, where his father was a chemical and dye merchant. Haber studied at Berlin and Heidelberg and then performed some research in organic chemistry. A period working for his father convinced him that he was not suited to a business career, and he moved to Jena to resume organic research. However, he was more interested in physical chemistry, and he soon moved to Karlsruhe to work on a study of the pyrolysis of hydrocarbons. Within a few years he had established a reputation as a leading physical chemist, his electrochemical work being especially noteworthy. In 1911 Haber accepted an invitation to head the newly established Kaiser Wilhelm Institute for Physical Chemistry and Electrochemistry near Berlin. At the outbreak of World War I Haber placed himself and the Institute at the service of the War Ministry. He was in scientific control of both the offensive and defensive aspects of Germany's chemical warfare programme. After the war he devised a method of extracting gold from sea water which he hoped would enable his country to pay the huge war reparations demanded by the Allies. The method failed because the concentration of gold in sea water was much lower than Haber had realised. Haber resigned his post and emigrated soon after the Nazis came to power in 1933 in protest against their anti–Semitic policies, and he died a few months later.

The growth of the textile industry in the Industrial Revolution created an increased demand for dyestuffs. Although the dyes in use at the time were natural in origin, dyeing processes used a number of chemical products, and dyeworks became important customers of the chemical industry. During the Industrial Revolution the chemical industry began to supply an increased range of inorganic chemicals for use as pigments. These were mainly incorporated into paints, but they were also used as food colorants, sometimes with disastrous consequences (Chapter 14). New bright colours for dyeing cloth became available from the middle of the nineteenth century, when the first synthetic organic dyes were introduced.

THE FIRST SYNTHETIC DYE

It is hardly surprising that the enormous advances in organic synthesis which took place after 1860 resulted in the discovery of compounds capable of commercial exploitation. The first such products were mainly dyes and drugs, and the industry which grew up to supply them became known as the fine chemicals industry. The first such compound to be marketed was discovered by accident in 1856 by William Henry Perkin (1838–1907) working under Hofmann's supervision at the Royal College of Chemistry in London.

Figure 15.1 Fritz Haber (1868–1934)

Perkin (Figure 15.2) was born in a poor part of East London where his father ran a successful carpentry and building business. Perkin demonstrated that he had some artistic talent, and his father wished him to train as an architect, no doubt with a view to employing him in the family business. Perkin himself had other ideas. He had become keenly interested in chemistry, and had set up his own laboratory at home. Perkin senior eventually agreed to send William to the Royal College of Chemistry to study under Hofmann. Although Perkin discontinued his chemical studies to build up his dyemaking business, he returned to research at the age of 36 when he sold his factory. He made important contributions to organic chemistry, among them the synthesis of the first perfume from coal tar (coumarin), and the synthesis of cinnamic acid by a method now known as the *Perkin reaction*.

Hofmann was investigating the compounds present in coal tar. This material had been available in large quantities since the introduction of coal gas in the first two decades of the nineteenth century, and was a useless waste product. While working with Liebig in Giessen, Hofmann had isolated aniline from coal tar, and he obtained benzene fairly soon after coming to London in 1845 to head the newly formed Royal College. One of Hofmann's students, Charles Blachford Mansfield (1819–1855), demonstrated that considerable quantities of benzene could be obtained by fractional distillation of the light oil obtained from coal tar. This fractionation process assumed great importance later in the century with the expansion of the fine chemicals industry. Tragically, Mansfield himself died when one of his stills boiled over and started a serious fire.

Figure 15.2 William Henry Perkin (1838–1907)

Perkin prepared a compound of composition $C_{10}H_{13}N$ (actually allyl toluidine). He hoped that by oxidation he might be able to produce the important drug quinine:

$$2C_{10}H_{13}N + 3O \rightarrow C_{20}H_{24}N_2O_2 + H_2O$$
quinine

The structural formulae of the organic compounds concerned were, of course, unknown at this stage, and so Perkin's suggested method of synthesis was not unreasonable. Perkin later recalled what happened when he treated allyl toluidine with the oxidising agent potassium dichromate:

> 'No quinine was formed, but only a dirty reddish–brown precipitate. Unpromising though this result was, I was interested in the action, and thought it desirable to treat a more simple base in the same manner. Aniline was selected, and its sulphate was treated with potassium dichromate; in this instance a black precipitate was obtained, and, on examination, this material was found to contain the colouring material since so well known as *aniline purple* or *mauve*'.

Perkin sent a sample of the new compound to a firm of dyers. Their assessment was very favourable, so at the early age of eighteen Perkin left the Royal College of Chemistry and raised sufficient capital to build a factory for the manufacture of mauve. Before mauve could be used as a dye, Perkin had to solve several technological problems connected with the dyeing process. His most significant innovation was the use of tannin as a mordant to bind the dye to the cotton.

THE GROWTH OF THE SYNTHETIC DYESTUFFS INDUSTRY

The opening of Perkin's factory in 1858 resulted in the formation of a fine chemicals industry which used coal tar as the raw material. The aniline which Perkin's process needed was obtained by the reduction of nitrobenzene, which in turn was made by the nitration of benzene obtained from coal tar. Perkin's discovery stimulated others to search for new compounds which might make good dyes. Very soon *magenta* had been produced by the oxidation of impure aniline (containing *ortho–* and *para*–toluidines). Hofmann showed that magenta was the salt of a base called rosaniline, and that different colours could be produced by introducing different alkyl groups into the molecule (which, like mauve, was of unknown structure).

In the year that Perkin's factory opened, Johann Peter Griess (1829–1888) found that aromatic amines were acted upon by nitrous acid, and that the resulting diazo compound could couple to another aromatic amine to produce a dye. In the 1870s it was discovered that diazo compounds would also couple to phenols producing yet another class of dye. These were initially known as *tropaeolines*.

All the dyestuffs mentioned so far had no natural counterparts. However, in 1869 Perkin in England and Heinrich Caro (1834–1910) in Germany independently discovered a method of synthesising alizarin, the natural dye of the madder plant, which had been used since antiquity. The starting material for the synthesis was the hydrocarbon anthracene. This is present in the least volatile fraction of coal tar, for which hitherto there had been very little use. Its value now increased enormously.

The advent of cheap synthetic alizarin resulted in the cessation of the cultivation of the madder plant. Some years later, most indigo plantations also became redundant. Natural indigo is a mixture obtained by fermentation and oxidation of the plant extract, and the principal coloured component is indigotin. Synthetic indigotin was available from the 1890s.

Consideration of the circumstances surrounding the introduction of mauve and indigotin shows how rapidly organic chemistry had changed in 30 years. Perkin had been trying to prepare quinine by a process which, with our modern knowledge of molecular structure, we can see was hopeless. The laboratory synthesis of indigotin was first achieved in 1880, and its structure was established in 1883 by von Baeyer, and this enabled rational attempts to be made to find a commercially viable synthesis.

The commercial production of dyes created a demand not only for the starting materials from coal tar but also for many organic reagents which were manufactured from the same source. There was also a demand for certain inorganic chemicals, for example sulphuric acid, sodium hydroxide and chlorine. Although the organic chemical industry had started in Britain, Germany soon became the world leader, partly because many more chemists were being trained in Germany as a result of the efforts of men like Liebig and Wöhler.

SYNTHETIC DRUGS

While synthetic dyes were the first organic chemicals to be manufactured in quantity, the branch of the chemical industry created to produce them was well placed to manufacture other useful organic compounds as they were discovered. Synthetic drugs started to make their appearance at the end of the nineteenth century, and probably the most famous of these was aspirin, introduced by the Bayer company in 1899. This and similar substances introduced about the same time were effective in alleviating symptoms but did nothing to cure the disease itself.

Compounds which would bind selectively to pathogenic microorganisms and destroy them were eagerly sought. A major success was achieved by Paul Ehrlich (1854–1915) in 1909, when he found that an organic arsenic compound was effective against the causative agent of syphilis. This compound was *Salvarsan*, although it was soon replaced by the closely related *Neosalvarsan*. These compounds are arsenic equivalents of azo dyes.

With the introduction of these anti–syphilitic agents, there were high hopes that chemists would soon develop compounds to combat other diseases. In fact, there were few important developments until 1935, when the German firm I. G. Farben patented a red dye under the name Prontosil. This was very effective against streptococcal and staphylococcal infections, but it was soon discovered by the husband and wife team J. and J. Trefouel, working at the Institut Pasteur in Paris, that Prontosil was broken down in the body to sulphanilamide, and that this was the genuine antibacterial agent. Sulphanilamide was a familiar compound that had been used for many years in the production of dyestuffs. Since it was not patentable, it rapidly became generally available. Within a few years over 5000 derivatives of sulphanilamide had been prepared and tested. Some of these *sulpha drugs* were found to be more effective against certain bacteria than sulphanilamide itself and found application in medical practice.

Valuable as the sulpha drugs were, their use was to a considerable extent superseded in the late 1940s and early 1950s by *antibiotics*. These are substances, produced by microorganisms, which are toxic to other microorganisms. This phenomenon, known as *antibiosis*, had first been observed by Pasteur in 1887, but all antibiotics investigated over the next 40 years seemed to be toxic to animal cells as well as to bacteria. Then in 1928 Alexander Fleming (1881–1928) observed that a Petri dish which had previously been overgrown with staphylococci had on standing become infected with a blue mould round which no staphylococci were now growing. It was subsequently found that the mould, *Penicillium notatum*, was producing an antibiotic quite unlike any which had been discovered previously. Penicillin was extraordinarily active against bacteria and, equally importantly, was non–toxic to animal tissues. The full evaluation of penicillin was not possible until it had been extracted and purified, and at the time this proved to be a task of insuperable difficulty.

In 1939 Howard Florey (1898–1968) and Ernst Chain (1906–1979), working in Oxford on a general study of the phenomenon of antibiosis, attempted once again to purify penicillin. Enough material was obtained by 1941 for an assessment of its remarkable properties to be made. Large–scale production was undertaken by several pharmaceutical companies in the United States, and by 1944 sufficient penicillin was available for use on Allied war casualties. In 1945 Fleming, Florey and Chain were awarded the Nobel Prize for Medicine. The synthesis of penicillin V was achieved by J. C. Sheehan and K. R. Henery–Logan in 1957, but the drug is still produced commercially by large–scale fermentation. Some penicillins currently in use have the side chain chemically modified after the precursor has been produced biosynthetically.

POLYMERS

Until World War II, the vast majority of synthetic organic chemicals produced were either dyes, drugs or intermediates used in their manufacture. Since World War II, not only have many more drugs (principally antibiotics) become available, but also a vastly increased range of organic compounds is manufactured for a multitude of purposes. Perhaps the most familiar category is polymers, and the most familiar applications of these compounds are as plastics and fibres. Although plastics have only been manufactured on a large scale since World War II, the plastics industry has its origins in the nineteenth century.

The first plastic materials to be utilised were natural materials, or modifications of them. As early as the seventeenth century, decorative mouldings were produced from the natural polymer, horn. In the eighteenth century, rubber and gutta percha, both gums from tropical trees, were first exploited. Around 1841 Charles Goodyear

(1800–1860) discovered that the properties of rubber could be much improved by heating with sulphur; this process, known as *vulcanisation*, was the first deliberate chemical modification of a natural polymer to produce a superior product.

The first plastic material to be made from non–plastic precursors was cellulose nitrate. This was obtained by Alexander Parkes (1813–1890) by treating cellulose fibres with nitric acid, and was first displayed at the Great International Exhibition in London in 1862 under the name *Parkesine*. Parkes moulded his new material into small decorative articles, as well as utilitarian objects such as knife handles. Parkesine was the first semi–synthetic plastic, so called because one of the starting materials was polymeric. The applications of cellulose nitrate were much extended by J. W. Hyatt (1837–1920) in the United States, who found that camphor was effective as a plasticiser, and the resulting mixture was known as *celluloid*. Another semi–synthetic plastic, cellulose acetate, was introduced around the end of the nineteenth century, and had the advantage over cellulose nitrate of being less flammable.

The first completely synthetic plastic material was made from the condensation of phenol and formaldehyde in the presence of a catalyst. The production of this material was perfected by Leo Hendrik Baekland (1863–1944), a Belgian chemist working in the United States, and it was marketed from 1909 under the name *Bakelite*. Bakelite is a highly crosslinked three–dimensional thermosetting polymer, and in the 1920s and 1930s a number of similar materials were developed such as urea formaldehyde and melamine formaldehyde.

The 1930s marked the start of the 'poly' era. It had been known since the nineteenth century that a number of organic compounds could yield tarry insoluble materials under certain conditions, but it was not until the 1930s that the chemical nature of these compounds was understood. It was Hermann Staudinger (1881–1965) who established that these materials, as well as natural rubber, consisted of very long–chain molecules. Commercial production of polyvinyl chloride (PVC) commenced in the 1930s with the development of suitable plasticisers, and other thermosoftening addition polymers which were first produced at this time include polystyrene, polyethylene (Polythene), polymethyl methacrylate (Perspex) and polychloroprene (Neoprene), which is similar to natural rubber except that the methyl side chain has been replaced by chlorine.

The chemist chiefly responsible for the development of Neoprene was Wallace Hume Carothers (1896–1937), working with the Du Pont Company in the United States. It was Carothers who was also responsible for the introduction of the first polyamide, nylon 66. It was after a careful analysis of the structure of silk that Carothers decided to synthesise polymeric amides. Nylon 66 was obtained by reacting hexamethylenediamine with adipic acid, and has been very successful as a synthetic fibre. Another class of linear condensation polymers which has found wide application as fibres is that of the polyesters.

Dyes, drugs and polymers are only three examples of the many different kinds of product manufactured by the modern fine chemicals industry. Others include anaesthetics, perfumes, detergents, herbicides, explosives, flavourings, preservatives and adhesives. Some of the reagents and solvents used in the manufacture of these materials are required in such quantity that their production is the basis of a sector of the industry now known as the heavy organic chemicals industry.

At the end of World War II the raw material for the manufacture of organic chemicals was still coal, but this has now been replaced by petroleum. The change occurred with the enormous increase in petroleum refining brought about by the extended use of the internal combustion engine, and with the simultaneous decrease in the manufacture of coal gas. The modern organic chemical industry is often called the petrochemical industry.

THE BENEFITS AND HAZARDS OF THE CHEMICAL AGE

To point out that we live in a world shaped to a very considerable extent by the labours of chemists and other scientists is to state the obvious. Nevertheless, this point needs emphasising when the problems associated with the chemical age are under discussion. Many of the advances of medical science that enable us to live longer result from the application of synthetic drugs. Pain can be controlled during surgery and serious illness by modern anaesthetics. Our food is protected during growth and storage by pesticides and preservatives. Much of our clothing is made from synthetic fibres, or a blend of synthetic and natural fibres. Plastics are now in evidence everywhere.

During the twentieth century, ways have been found to increase agricultural production enormously. This has been chiefly due to the researches of chemists and of plant breeders, and between 1950 and 1984 world grain production per person increased by nearly 40 per cent. Since then population increase has outstripped further gains in agricultural production. Although the chemical industry has so far helped to prevent worldwide starvation, it is clear that there is a desperate need for a massive international effort by politicians, economists and scientists to stabilise the population of the planet and to provide for the needs of everyone. Such a programme will be enormously expensive, but the consequences of inaction would be catastrophic.

It is important to remember that chemistry, like any other science, is capable of sinister applications as well as beneficial ones. If chemistry is to take the credit for making antibiotics available, it cannot shirk the blame for the development of chemical weapons. However, many scientific and technological advances present opportunities for evil uses as well as beneficial ones, and once a particular piece of research has been published the original author has no control over its application.

Before World War II, most of the chemicals manufactured in large quantity were used to make products such as glass, soap, paper, fertilisers, metals, and so on. Today, we all come into direct contact with a much larger number of products of the chemical industry, and this has heightened concern about the safety of some of these substances. After tragedies such as that which ensued when the drug Thalidomide was taken by some pregnant women, the testing and quality control of chemicals intended for human consumption have become more rigorous, and a number of drugs, food additives and flavourings have been removed from the market because of possible health hazards.

There have also been disasters associated with the manufacture of chemicals. The most notorious was the accident at Bhopal in India in 1984, which resulted in approximately 3000 deaths when leakage occurred from an agrochemical plant. Some workers in chemical factories have contracted occupational diseases as a result of exposure to chemicals. A well known example was the high incidence of bladder cancer among workers coming into contact with bicyclic aromatic amines such as 2–naphthylamine.

Pollution caused by effluents from chemical works is not a new problem, and the discharges emanating from some of the earlier chemical factories (for example the Leblanc alkali works) were quite horrendous by modern standards. One example of a twentieth century pollution problem is provided by Minamata disease. This was caused by the discharge of mercury, used as a catalyst in a plastics factory, into Minamata Bay in Japan. Some of those who ate fish caught in the bay suffered damage to the nervous system, and some children born to mothers who had eaten the fish had serious birth defects. Although mercury was identified as the cause of the problem in 1953, it was not understood how such low concentrations of the metal could cause such serious problems. It was only in 1969 that it was realised that the real culprit was dimethylmercury, which had been formed from metallic mercury by bacteria in the mud at the bottom of the bay.

Apart from problems caused by effluents and by accidents at chemical factories, the use of some chemical products has had unforeseen effects on the environment. An example is provided by the chlorinated hydrocarbon insecticides. Dichlorodiphenyltrichloroethane (DDT) was first prepared in 1874, and in 1936 it was found to be a very powerful insecticide and was used with great success against mosquitoes, flies, lice and other vectors of disease, and also against many agricultural pests. Other useful compounds were soon developed which were also chlorinated hydrocarbons. Among these were hexachlorocyclohexane (known as BHC from its alternative name of benzene hexachloride), heptachlor, aldrin and dieldrin. The last two were so named because they were synthesised by Diels–Alder reactions. Although these insecticides were extremely effective, it became apparent that there were serious disadvantages associated with their use. They were found to concentrate in the fatty tissues of animals and birds preying on insects which had received non–lethal doses. The insecticides thus entered the food chain, and in some cases were the cause of dramatic reductions in bird populations. The use of chlorinated hydrocarbon insecticides is now restricted, and other pesticides which were formerly permitted have been withdrawn because of the persistence of their residues.

Another example of problems resulting from the release of chemicals into the environment is displayed by the chlorofluorocarbons (CFCs). These very stable compounds, which have been widely used as refrigerants and aerosol propellants, are believed to be damaging the atmospheric ozone layer. They are now being replaced by more 'ozone–friendly' products.

In the second half of the twentieth century, humankind has enjoyed the enormous benefits of the chemical age, but has also learned by bitter experience of the problems that can occur as a result of the manufacture and use of certain chemical products. We should now have enough experience to continue to enjoy these benefits but to minimise the chance of future disasters. Not only must we continue to set rigorous standards for chemical products, but chemical factories must also be built to very high specifications and must be subject to the most stringent controls regarding their operating procedures and any effluent they discharge. These controls should be agreed internationally to prevent manufacturers in one country being placed at a disadvantage by others elsewhere in the world where standards might be less strict or less rigorously enforced.

It would be naive to expect that no more problems will arise in the future. In spite of all our efforts, the introduction of new products and processes may still have unforeseen consequences. However, such difficulties will only be overcome in the way that similar problems have been resolved in the past. This is by more research, leading to safer and more effective substances and better techniques for their manufacture.

The first chemical reaction to be exploited by man was that of combustion. There is a certain irony in the fact that one of the products of the combustion of fuels, namely carbon dioxide, should be the cause of a possible threat to the future of the planet. If worries about the greenhouse effect prove to be well founded, the only chance of averting serious global warming will be by further research to develop alternative energy sources. Just as the widespread famine about which William Crookes warned at the end of the nineteenth century was prevented by the invention of the Haber process, so contemporary fears can only be allayed by further advances in chemistry and in other branches of science and technology. One thing is certain; there can be no turning back.

Appendix

The Nobel Prize in Chemistry

Alfred Bernard Nobel (1833–1896) amassed an enormous fortune from his inventions and improvements in the manufacture of explosives. His father was also an explosives manufacturer, and in 1863 Alfred developed a detonator based on mercury fulminate, which made possible the use of the liquid explosive nitroglycerine. Nobel continued his experiments in spite of an explosion in 1864 that destroyed the factory and killed five people including his younger brother. In 1867 he patented *dynamite*, in which nitroglycerine was absorbed by the inert solid kieselguhr and was therefore much safer to handle. In 1875 he introduced the more powerful *blasting gelatine*, in which the nitroglycerine was gelatinised with nitrocellulose. These inventions made possible major civil engineering projects like the Corinth canal and the St Gotthard tunnel. In 1887 Nobel introduced *ballistite*, a smokeless explosive for military use. Nobel hoped that the destructive capabilities of the new explosives would reduce the likelihood of war.

Nobel left his fortune for the establishment of five prizes to be awarded annually for achievements in chemistry, physics, physiology or medicine, literature of an idealistic tendency, and the promotion of world peace. The first awards were made in 1901. The Nobel Prize for Economics was founded in 1968 by the National Bank of Sweden and the first award was made in 1969. The Nobel Prizes have become the most highly regarded of all international awards. A Prize cannot be shared by more than three people, and cannot be awarded posthumously. A list of the winners of the Nobel Prize for Chemistry is given below. On occasions the work that has resulted in awards in physics and physiology or medicine has been to some extent chemical in content or application.

1901	Jacobus Henricus van't Hoff (1852–1911)	'For the laws of chemical dynamics and of osmotic pressure.'
1902	Emil Hermann Fischer (1852–1919)	'For his syntheses in the groups of sugars and purines.'
1903	Svante August Arrhenius (1859–1927)	'For his theory of electrolytic dissociation.'

1904 William Ramsay
(1852–1916)

'For the discovery of the gaseous, indifferent elements in the air and the determination of their place in the periodic system.'

1905 Johann Friedrich Wilhelm Adolf von Baeyer
(1835–1917)

'For his researches on organic dyestuffs and hydroaromatic compounds.'

1906 Ferdinand–Frédérick Henri Moissan
(1852–1907)

'For his investigation and isolation of the element fluorine and for placing at the service of science the electric furnace which bears his name.'

1907 Eduard Buchner
(1860–1917)

'For his biochemical researches and his discovery of cell–free fermentation.'

1908 Ernest Rutherford
(1871–1937)

'For his investigations into the disintegration of the elements and the chemistry of radioactive substances.'

1909 Friedrich Wilhelm Ostwald
(1853–1932)

'For his work on catalysis and for his investigations into the fundamental principles governing chemical equilibria and rates of reaction.'

1910 Otto Wallach
(1847–1931)

'For his pioneer work in the field of alicyclic compounds.'

1911 Marie Sklodowska Curie
(1867–1934)

'For her services to the advancement of chemistry by the discovery of the elements radium and polonium, by the isolation of radium and the study of the nature and compounds of this remarkable element.'

1912 Françoise August Victor Grignard
(1871–1935)

'For the discovery of the so–called Grignard reagent, which in recent years has greatly advanced the progress of organic chemistry.'

Paul Sabatier
(1854–1941)

'For his method of hydrogenating organic compounds in the presence of finely divided metals.'

1913 Alfred Werner
(1866–1919)

'In recognition of his work on the linkage of atoms in molecules, by which he has thrown fresh light on old problems and opened up new fields of research, especially in inorganic chemistry.'

1914 Theodore William Richards
(1868–1928)

'For his exact determination of the atomic weights of a large number of chemical elements.'

1915 Richard Martin Willstätter
(1872–1942)

'For his researches on plant pigments, especially chlorophyll.'

1916 No award

1917 No award

1918 Fritz Haber
(1868–1934)

'For the synthesis of ammonia from its elements.'

1919 No award

1920 Hermann Walther Nernst
(1864–1941)

'In recognition of his work in thermochemistry.'

1921 Frederick Soddy
(1877–1956)

'For his important contributions to our knowledge of the chemistry of radioactive substances and his investigations into the origin and nature of isotopes.'

1922 Francis William Aston
(1877–1945)

'For his discovery, by means of his mass spectrograph, of isotopes in a large number of non–radioactive elements, and for his enunciation of the whole–number rule.'

1923 Fritz Pregl
(1869–1930)

'For his invention of the method of microanalysis of organic substances.'

1924   No award

1925   Richard Adolf Zsigmondy
       (1865–1929)

'For his demonstration of the
heterogeneous nature of colloid
solutions and for the methods he
used, which have since become
fundamental in modern colloid
chemistry.'

1926   Theodor Svedberg
       (1884–1971)

'For his work on disperse systems.'

1927   Heinrich Otto Wieland
       (1877–1971)

'For his investigations of the
constitution of the bile acids
and related substances.'

1928   Adolf Otto Reinhold Windaus
       (1876–1959)

'For the services rendered through
his research into the constitution
of the sterols and their
connection with the vitamins.'

1929   Arthur Harden
       (1865–1940)
       Hans Karl August Simon
       von Euler–Chelpin
       (1873–1964)

'For their investigations of the
fermentation of sugar and
fermentative enzymes.'

1930   Hans Fischer
       (1881–1945)

'For his researches into the
constitution of hemin and
chlorophyll and especially for
his synthesis of hemin.'

1931   Carl Bosch
       (1874–1940)
       Friedrich Bergius
       (1884–1949)

'In recognition of their
contributions to the invention and
development of chemical high–
pressure methods.'

1932   Irving Langmuir
       (1881–1957)

'For his outstanding discoveries
and investigations within the
field of surface chemistry.'

1933   No award

1934   Harold Clayton Urey
       (1893–1981)

'For his discovery of heavy
hydrogen.'

1935　Irène Joliot–Curie
　　　(1897–1956)
　　　Jean Frédéric Joliot
　　　(1900–1958)

'For their synthesis of new radioactive elements.'

1936　Peter Joseph William Debye
　　　(1884–1966)

'For his contributions to the study of molecular structure through his investigations on dipole moments and on the diffraction of X–rays and electrons in gases.'

1937　Walter Norman Haworth
　　　(1883–1950)

'For his investigations on carbohydrates and vitamin C.'

　　　Paul Karrer
　　　(1889–1971)

'For his researches and investigations on carotinoids, flavins, and vitamins A and B.'

1938　Richard Kuhn
　　　(1900–1967)

'For his work on carotinoids and vitamins.'

1939　Adolf Butenandt
　　　(b. 1903)

'For his work on the mammalian sex hormones.'

　　　Leopold Stephen Ruzicka
　　　(1887–1976)

'For his work on polymethylenes and higher terpenes.'

1940　No award

1941　No award

1942　No award

1943　George Charles Hevesy
　　　(1885–1966)

'For his work on the use of isotopes as tracers in the study of chemical processes.'

1944　Otto Hahn
　　　(1879–1968)

'For his discovery of the fission of heavy nuclei.'

1945　Artturi Ilmari Virtanen
　　　(1895–1973)

'For his research and discoveries in the field of agricultural and nutrition chemistry, and particularly his method of preserving animal fodder.'

1946   James Batcheller Sumner (1887–1955)

'For his discovery that enzymes can be crystallised.'

John Howard Northrop (1891–1987)
Wendell Meredith Stanley (1904–1971)

'For their preparation of enzymes and virus proteins in pure form.'

1947   Robert Robinson (1886–1975)

'For his investigations on plant products of biological importance, especially the alkaloids.'

1948   Arne Wilhelm Kaurin Tiselius (1902–1975)

'For his research on electrophoresis and on analysis by adsorption, in particular for his discoveries concerning the heterogeneous nature of the proteins of the serum.'

1949   William Francis Giauque (1895–1982)

'For his contributions in the field of chemical thermodynamics, particularly concerning the behaviour of substances at extremely low temperatures.'

1950   Otto Paul Hermann Diels (1876–1954)
Kurt Alder (1902–1958)

'For their discovery and development of the diene synthesis.'

1951   Glenn Theodore Seaborg (*b*. 1912)
Edwin Mattison McMillan (*b*. 1907)

'For their discoveries in the chemistry of the transuranium elements.'

1952   Archer John Porter Martin (*b*. 1910)
Richard Laurence Millington Synge (*b*. 1914)

'For their invention of partition chromatography.'

1953   Hermann Staudinger (1881–1965)

'For his discoveries in the field of macromolecular chemistry.'

1954	Linus Pauling (b. 1901)	'For his research into the nature of the chemical bond and its application to the structure of complex substances.'
1955	Vincent du Vigneaud (1901–1978)	'For his work on biochemically important sulphur compounds and particularly for the first synthesis of a polypeptide hormone.'
1956	Cyril Norman Hinshelwood (1897–1967) Nikolai Nikolaevich Semenov (1896–1986)	'For their researches into the mechanisms of chemical reactions.'
1957	Alexander Robertus Todd (b. 1907)	'For his work on nucleotides and nucleotide coenzymes.'
1958	Frederick Sanger (b. 1918)	'For his work on the structure of proteins, especially that of insulin.'
1959	Jaroslav Heyrovsky (1890–1967)	'For inventing and developing the polarographic method of analysis.'
1960	Willard Frank Libby (1908–1980)	'For his method to use carbon 14 for age determination in archaeology, geology, geophysics and other branches of science.'
1961	Melvin Calvin (b. 1911)	'For his work on the carbon dioxide assimilation in plants.'
1962	John Cowdery Kendrew (b. 1917) Max Ferdinand Perutz (b. 1914)	'For their studies of the structure of globular proteins.'
1963	Guilio Natta (1903–1979) Karl Ziegler (1898–1973)	'For their discoveries in the field of the chemistry and technology of high polymers.'

1964 Dorothy Mary Crowfoot Hodgkin
(*b*. 1910)

'For her determination by X–ray techniques of the structures of important biochemical substances.'

1965 Robert Burns Woodward
(1917–1979)

'For his outstanding achievements in the art of organic synthesis.'

1966 Robert Sanderson Mulliken
(1896–1986)

'For his fundamental work concerning chemical bonds and the electronic structure of molecules by the molecular orbital method.'

1967 Manfred Eigen
(*b*. 1927)
Ronald George Wreyford Norrish
(1897–1978)
George Porter
(*b*. 1920)

'For their studies of extremely fast chemical reactions, effected by disturbing the equilibrium by means of very short pulses of energy.'

1968 Lars Onsager
(1903–1976)

'For the discovery of the reciprocal relations bearing his name, which are fundamental for the thermodynamics of irreversible processes.'

1969 Derek Harold Richard Barton
(*b*. 1918)
Odd Hassel
(1897–1981)

'For their contributions to the development of the concept of conformation and its application in chemistry.'

1970 Luis Federico Leloir
(*b*. 1906)

'For his discovery of sugar nucleotides and their role in the biosynthesis of carbohydrates.'

1971 Gerhard Herzberg
(*b*. 1904)

'For his contributions to the knowledge of electronic structure and geometry of molecules, particularly free radicals.'

1972 Christian Boehmer Anfinsen
(*b*. 1916)

'For his work on ribonuclease, especially concerning the connection between the amino acid sequence and the biologically active conformation.'

	Stanford Moore (1913–1982) William Howard Stein (1911–1980)	'For their contribution to the understanding of the connection between chemical structure and catalytic activity of the active centre of the ribonuclease molecule.'
1973	Ernst Otto Fischer (*b.* 1918) Geoffrey Wilkinson (*b.* 1921)	'For their pioneering work, performed independently, on the chemistry of organometallic, so–called sandwich compounds.'
1974	Paul John Flory (1910–1985)	'For his fundamental achievements, both theoretical and experimental, in the physical chemistry of the macromolecules.'
1975	John Warcup Cornforth (*b.* 1917)	'For his work on the stereo-chemistry of enzyme–catalysed reactions.'
	Vladimir Prelog (*b.* 1906)	'For his research into the stereochemistry of organic molecules and reactions.'
1976	William Nunn Lipscomb (*b.* 1919)	'For his studies on the structure of boranes illuminating the problems of chemical bonding.'
1977	Ilya Prigogine (*b.* 1917)	'For his contributions to nonequilibrium thermodynamics, particularly the theory of dissipative structures.'
1978	Peter Dennis Mitchell (*b.* 1920)	'For his contribution to the understanding of biological energy transfer through the formulation of the chemiosmotic theory.'
1979	Herbert Charles Brown (*b.* 1912) Georg Friedrich Karl Wittig (*b.* 1897–1987)	'For their development of boron- and phosphorus–containing compounds, respectively, into important reagents in organic synthesis.'

1980   Paul Berg
       (*b*. 1926)

'For his fundamental studies of the biochemistry of nucleic acids, with particular regard to recombinant DNA.'

Walter Gilbert
(*b*. 1932)
Frederick Sanger
(*b*. 1918)

'For their contributions concerning the determination of base sequences in nucleic acids.'

1981   Kenichi Fukui
       (*b*. 1918)
       Roald Hoffmann
       (*b*. 1937)

'For their theories, developed independently, concerning the course of chemical reactions.'

1982   Aaron Klug
       (*b*. 1926)

'For his development of crystallographic electron microscopy and his structural elucidation of biologically important nucleic acid–protein complexes.'

1983   Henry Taube
       (*b*. 1915)

'For his studies of the mechanisms of electron transfer reactions, particularly of metal complexes.'

1984   Robert Bruce Merrifield
       (*b*. 1921)

'In recognition of his methodology for chemical synthesis on a solid matrix.'

1985   Herbert Aaron Hauptmann
       (*b*. 1917)
       Jerome Karle
       (*b*. 1917)

'For their outstanding achievements in the development of direct methods for the determination of crystal structures.'

1986   Dudley Robert Herschbach
       (*b*. 1932)
       Yuan Tseh Lee
       (*b*. 1936)
       John Charles Polanyi
       (*b*. 1929)

'For their contributions concerning the dynamics of elementary chemical processes.'

1987   Charles Pedersen
       (1904–1988)
       Donald Cram
       (*b.* 1919)
       Jean–Marie Lehn
       (*b.* 1939)

'For their development and use of molecules with structure–specific interactions of high selectivity.'

1988   Johann Deisenhofer
       (*b.* 1943)
       Robert Huber
       (*b.* 1937)
       Hartmut Michel
       (*b.* 1946)

'For their work in the field of photosynthesis.'

1989   Sidney Altman
       (*b.* 1939)
       Thomas Cech
       (*b.* 1948)

'For their discovery of the catalytic function of the genetic material RNA.'

1990   Elias James Corey
       (*b.* 1928)

'For finding new ways of producing and synthesising chemical compounds.'

1991   Richard Ernst
       (*b.* 1933)

'For the development of nuclear magnetic resonance spectroscopy.'

1992   Rudolph Marcus
       (*b.* 1923)

'For his theoretical work on electron transfer between molecules.'

Bibliography

A considerable number of journals deal with the history of science and medicine, but only one (*Ambix*, published by the Society for the History of Alchemy and Chemistry) is devoted specifically to the history of chemistry. Articles of historical interest frequently appear in journals such as *Chemistry in Britain*, *Journal of Chemical Education* and *Education in Chemistry*. Historical articles also appear occasionally in more general journals such as *Scientific American* and *New Scientist*, as well as in specialised journals concerned with particular subdivisions of chemistry. An excellent guide to the literature published in the period from approximately 1960 to 1980 is to be found in C. A. Russell (ed.), *Recent Developments in the History of Chemistry* (London, 1985). Anyone wishing to explore an aspect of the history of chemistry in greater depth would be well advised to start with that publication.

The book list that follows is inevitably somewhat idiosyncratic, but it comprises a selection of some of the more readily available works that the author has found to be of interest and value. Included in the list is a selection of reprints of famous books and papers. Those seriously interested in the history of chemistry should read a selection of the classics of the subject, and some of the publications listed reproduce the original works in facsimile. While few of us may be able to consult a contemporary copy of Robert Kerr's translation of Lavoisier's *Traité* published in 1790, a facsimile edition is readily available, and we can thus attempt to imagine ourselves as one of the original readers of that great classic of chemistry.

General Works

E. Farber (ed.), *Great Chemists* (Interscience: New York, 1961).

A. Findlay, *A Hundred Years of Chemistry*, 3rd edn. rev. T. I. Williams (Methuen: London, 1965).

C. C. Gillispie (ed.). *Dictionary of Scientific Biography* (Charles Scribner's Sons: New York, 1970–80).

A. J. Ihde, *The Development of Modern Chemistry* (Dover: New York, 1984).

H. M. Leicester, *The Historical Background of Chemistry* (Dover: New York, 1971).

J. R. Partington, *A History of Chemistry*, 4 vols (Macmillan: London, 1961–70).

J. R. Partington, *A Short History of Chemistry*, 2nd edn. (Macmillan: London, 1948).

C. A. Russell (ed.), *Recent Developments in the History of Chemistry* (Royal Society of Chemistry: London, 1985).

Books on Particular Topics

W. H. Brock, *From Protyle to Proton* (Hilger: Bristol, 1985).
W. A. Campbell, *The Chemical Industry* (Longmans: London, 1971).
D. S. L. Cardwell (ed.), *John Dalton and the Progress of Science* (Manchester University Press, Manchester, 1968).
M. P. Crosland, *Historical Studies in the Language of Chemistry*, 2nd edn. (Dover: New York, 1978).
E. J. Holmyard, *Alchemy* (Dover: New York, 1990).
H. M. N. H. Irving, *The Techniques of Analytical Chemistry* (HMSO: London, 1974).
G. B. Kauffman, *Inorganic Coordination Compounds* (Heyden: London, 1981).
H. Kearney, *Science and Change* (Weidenfeld and Nicholson: London, 1971).
G. E. R. Lloyd, *Early Greek Science: Thales to Aristotle* (Chatto & Windus: London, 1970).
G. E. R. Lloyd, *Greek Science after Aristotle* (Norton: New York, 1973).
D. McKie, *Antoine Lavoisier, Scientist, Economist, Social Reformer* (Constable: London, 1952).
C. A. Russell, *The History of Valency* (Leicester University Press: Leicester, 1971).
F. Szabadváry, *History of Analytical Chemistry* (Pergamon: Oxford, 1966).
F. S. Taylor, *The Alchemists* (Heinemann: London, 1951).
T. Wasson (ed.), *Nobel Prize Winners* (Wilson: New York, 1987).
M. E. Weeks and H. M. Leicester, *Discovery of the Elements*, 7th edn. (Journal of Chemical Education: Easton, Pennsylvania, 1968).
S. Weinberg, *The Discovery of Subatomic Particles* (Scientific American Books: New York, 1983).
T. I. Williams, *The Chemical Industry*, 2nd edn. (EP Publishing: Wakefield, 1972).

Reprints of Famous Books and Papers

J. Dalton, *A New System of Chemical Philosophy* (Peter Owen: London, 1965).
R. Kirwan, *An Essay on Phlogiston* (Frank Cass: London, 1968).
D. M. Knight (ed.), *Classical Scientific Papers, Chemistry* (Mills & Boon: London, 1968).
D. M. Knight (ed.), *Classical Scientific Papers, Chemistry*, Second Series (Mills & Boon: London, 1970).
A. L. Lavoisier, *Elements of Chemistry*, transl. Robert Kerr (Dover: New York, 1965).
H. M. Leicester and H. S. Klickstein (eds), *Source Book in Chemistry* 1400–1900 (McGraw–Hill: New York, 1952).

H. M. Leicester (ed.), *Source Book in Chemistry*, 1900–1950 (Harvard University Press: Cambridge, Massachusetts, 1968).

Name Index

Subject Index